SpringerBriefs on Cyber Security Systems and Networks

Editor-in-Chief

Yang Xiang, Digital Research and Innovation Capability, Swinburne University of Technology, Hawthorn, Melbourne, VIC, Australia

Series editors

Liqun Chen, University of Surrey, Guildford, UK
Kim-Kwang Raymond Choo, University of Texas at San Antonio, San Antonio, TX, USA
Sherman S. M. Chow, Department of Information Engineering, The Chinese University of Hong Kong, Shatin, Hong Kong
Robert H. Deng, School of Information Systems, Singapore Management University, Singapore, Singapore
Dieter Gollmann, Hamburg University of Technology, Hamburg, Germany
Javier Lopez, University of Málaga, Málaga, Spain
Kui Ren, University at Buffalo, Buffalo, NY, USA
Jianying Zhou, Singapore University of Technology and Design, Singapore, Singapore

The series aims to develop and disseminate an understanding of innovations, paradigms, techniques, and technologies in the contexts of cyber security systems and networks related research and studies. It publishes thorough and cohesive overviews of state-of-the-art topics in cyber security, as well as sophisticated techniques, original research presentations and in-depth case studies in cyber systems and networks. The series also provides a single point of coverage of advanced and timely emerging topics as well as a forum for core concepts that may not have reached a level of maturity to warrant a comprehensive textbook. It addresses security, privacy, availability, and dependability issues for cyber systems and networks, and welcomes emerging technologies, such as artificial intelligence, cloud computing, cyber physical systems, and big data analytics related to cyber security research. The mainly focuses on the following research topics:

Fundamentals and Theories

- Cryptography for cyber security
- Theories of cyber security
- Provable security

Cyber Systems and Networks

- Cyber systems security
- Network security
- Security services
- Social networks security and privacy
- Cyber attacks and defense
- Data-driven cyber security
- Trusted computing and systems

Applications and Others

- Hardware and device security
- Cyber application security
- Human and social aspects of cyber security

More information about this series at http://www.springer.com/series/15797

Chee Keong Ng · Lei Pan
Yang Xiang

Honeypot Frameworks and their Applications: A New Framework

 Springer

Chee Keong Ng
School of Information
Deakin University
Burwood, Melbourne, VIC
Australia

Yang Xiang
Digital Research and Innovation Capability
Swinburne University of Technology
Hawthorn, Melbourne, VIC
Australia

Lei Pan
School of Information
Deakin University
Burwood, Melbourne, VIC
Australia

ISSN 2522-5561   ISSN 2522-557X   (electronic)
SpringerBriefs on Cyber Security Systems and Networks
ISBN 978-981-10-7738-8   ISBN 978-981-10-7739-5   (eBook)
https://doi.org/10.1007/978-981-10-7739-5

Library of Congress Control Number: 2018938790

Printed on acid-free paper

This Springer imprint is published by the registered company Springer Nature Singapore Pte Ltd. part of Springer Nature
The registered company address is: 152 Beach Road, #21-01/04 Gateway East, Singapore 189721, Singapore

This book is dedicated to those who are interested to know more about honeypots. It does not matter whether you are an expert or novice, this book is for you.

Preface

Most people understand honeypots as systems sit in an isolated corner of the network waiting for attacker to discover and compromise them. This is often untrue, in fact, in some frameworks, honeypots have enjoyed the prime spot in an organisational network to lure potential hacker. As new instances of malware appear so rapidly that more spotlight has been placed in honeypot technology.

This book gives a detailed description of honeypots including their forms, purposes, natures and interaction. It also gives an in-depth introduction of different types of honeypot, their applications in monitoring and capturing of malware and adversary tactic to detect honeypot.

The main role of honeypot which effectively assists the researcher to derive solutions for the deadly malware attack has been outlined in the book. This book also gives rich information of other roles and uses of honeypot not only in the area of cyber and network security, but also in collecting proof for forensic investigation.

Finally, this book addresses the importance of honeypot as a learning tool for detecting future malware such as ransomware.

Burwood, Melbourne, Australia   Chee Keong Ng
Burwood, Melbourne, Australia   Lei Pan
Hawthorn, Melbourne, Australia   Yang Xiang

Acknowledgements

Thanks to Almighty God for enabling and helping me to complete this book.

I would like to first express my appreciation to my family, especially my wife, Xinying Liu, for her support and love. I would like to also express my most sincere gratitude to my supervisor, Prof. Xiang Yang, and Dr. Lei Pan for their wisdom and advice which is essential for the completion of this book.

Next, I would like to also express my token of appreciation to NSCLab, especially Dr. Jun Zhang for their moral support.

Lastly, I would like to thank Springer Publication for giving me this opportunity to spare my knowledge and finding in their book. I would also like to express my thanks to the editor, Dr. Xiaolan Yao, for her patient and assistance.

Contents

Chapter 1
Introduction to Honeypot

Abstract Honeypot is a decoy system or a simulated application which simulates an entire network to lure attacker by disguising itself with popular vulnerabilities. There are different types of honeypots. For instance, a research honeypot can assist researchers to monitor and analyse the activities of the attacker that are captured in the honeypot. Usually, honeypot can be categorised into three different sub-types based on its purpose, interaction and form. It is then further categorised according to its nature, specialization and framework. Honeypot, however, is not a foolproof concept; often it can be detected by experienced attacker. The information about the features of honeypot and anti-honeypot tools are widely available online to educate attackers. This book will cover the honeypot to detect some of the more popular and damaging attacks such as worm, DDoS, APT, phishing and insider breaches. It will also cover the application of forensics work in honeypot and proposed concept from honeypot researchers to enhance the features of honeypot so as to make it difficult distinguish between a real host and honeypot.

1.1 Introduction

When honeypot was introduced by Fred Cohen in 1998 [1], it was known as a deception toolkit which is used to trap attacker with a large number of widely known vulnerabilities. The term "honeypot" was first used by Lance Spitzer in 2002 to describe how attacker is lurked toward a system perceive to possess valuable information similar to the scenario when a bee is attracted to the honey in the honeypot [2]. The initial intention of honeypot is for research purpose. It is used to capture, analyze, and derive the motivation behind an attack. Different types of honeypots, such as honeyd [3] a limited interactive honeypot and Sebek [4] real system with real operating system honeypot, are created by researchers which enables honeypots to interact with the attacker to improve the concealment of the honeypot. Honeypot has evolved from merely a toolkit into a real system equipped with "real" data.

Honeypots have been divided into different categories. Most people understand the difference of honeypot based on their capability of interaction such as the low

C. K. NG et al., *Honeypot Frameworks and their Applications:*
A New Framework, SpringerBriefs on Cyber Security Systems and Networks,
https://doi.org/10.1007/978-981-10-7739-5_1

interaction honeypot and high-interaction honeypot. The distinctions of honeypots can also be categorised based on their form type, purpose used, structure and nature. Honeypots can be used to cater for a general attack such as server-based attack and client-based attack, or more specialized attack such as APT [5], worm, DDoS [6] and insider [7].

The importance of the honeypot has increased rapidly, especially in the early part of this millennium where worms, botnet and other malware attacks have become more frequent and common causing serious losses economically. Researchers have proposed various honeypot frameworks to tackle the ever-growing issue. This includes the application of:

1. The multiple honeynet framework for polymorphic worm [8–11]
2. The usage of the advantage of low and high interactive honeypot to capture malware exploitation [12, 13]
3. Honeypot with redirect toolkit or anomaly detection to capture botware [14–16]

Honeypots have also being implemented as part of the security system in a production environment serve as an early warning to any intrusion discovered. This is to distract the attacker away from the real system. Even though honeypots can be used to prevent attack, its main responsibilities such as monitoring, collecting and analyzing the activities of the attacker and attack remain unchanged but to a much lesser degree. While the work of honeypot continues, honeypots do contribute greatly in the work of forensic analysis. A new term "honeypot forensic" refers to the application of forensic work in the honeypot. Some researchers [17] believe that the traditional forensics method has become irrelevant and is no longer adequate to meet its expectation when analyzing the data collected using honeypots.

Honeypots do face several technical challenges. Honeypots do have some fixed identities/features which will give themselves away. This is also due to the availability of information about honeypot online such as Project Honeypot, www.project-honeypot.net and honeynet.org [18]. These sites provide the downloads for all open-source honeypot application and source code. The attacker can easily learn and study about the strength and weakness of honeypots.

Attackers can use methods such as fingerprint attack [19], the response timing analysis [20, 21] and send attack traffic from the compromised machine [22] to determine the presence of honeypot. There are also many open source anti-honeypot toolkits such as honeypot hunter which is freely available to be used to detect honeypot.

Honeypots also face other challenges such as honeypot immersion [23]. Honeypot immersion describes the experience of adversaries when engaging with the honeypot system. Sometime the ease of compromising of a honeypot may held back or scare off a more experienced person from attacking the honeypot.

Honeypot has been a very useful tool to learn about ransomware and its attack. However, there are very little paper about the application of honeypot in ransomware attack, despite the growing popularity of ransomware. Ransomware encrypts file of legitimate users and demands an extortsion from them [24]. It demands payment in form of cryptocurrency such as bitcoin. Ransomware has caused devastating damage financially to the society and is the fastest growing malware. The variants of the ransomware has increase ten-fold since 2013. Client honeypot equipped with CaptureBat [25] collects the information of amended registry, file activities and the communication between ransomware and its CNC server. The information collected is used for these two purposes, to formula a more effective signature and to derive a solution to decrypt the encryption. The proposed solution will use multiple server-based honeypots which equipped with cuckoo sandbox [26] and remunx framework [27], and signature generator to enhance its efficiency to capture the full capabilities of the ransomware.

Honeypot is used to capture a specific pattern of different types of ransomware. Honeypot using honeytoken is used to trace the bitcoin transaction to its destination. Signal beacon is embedded into the bitcoin transaction so that once the transaction reached its destination and accessed [28]. The beacon will be activated and send back to the sensor server.

The aim of this survey are to:

1. Demonstrate the great flexibility of honeypots
2. Provides rich inside for its reader to understand different honeypot approach to the same problem
3. Outline the limitation of honeypot and solutions to their limitations
4. Present current anti-honeypot methods and their counter method
5. Contribute honeypot and honeytoken to popular malware such as ransomware

The rest of the book are organized as follows. Chapter 2 explains the basic and advance concept of honeypot and the roadmap of the whole survey. Chapters 3, 4 and 5 describes the difference ground breaking concept of honeypots proposed by researcher and their uses. Chapter 3 will described the specialized client and server honeypot in great length. Chapter 4 focuses on general purposed honeypot include shadow honeypot and dynamic honeypot. The various method of cover and concealment that contribute to the succeed of reconnaissance work performed by the honeypot will be covered in Chap. 5. The use of honeypot in the area of forensic work will be discussed at the end of Chap. 5. Chapter 6 summarizes the conceptual framework of the honeypot. It also describes the anti-honeypot methods and their proposed counter-measures. Chapter 7 discusses the application of honeypot concept on ramsonware and followed by Chap. 8 which concludes the book and sums up the future research work.

References

1. F. Cohen et al., The deception toolkit. Risks Digest **19**, 1998 (1998)
2. L. Spitzner, Honeypots: catching the insider threat, in *19th Annual on Computer Security Applications Conference, 2003. Proceedings* (IEEE, 2003), pp. 170–179

3. N. Provos, Honeyd-a virtual honeypot daemon, in *10th DFN-CERT Workshop*, vol. 2 (Hamburg, Germany, 2003), p. 4
4. K.Y. Enemy, Sebek, a kernel based data capture tool, the honeynet project (2003)
5. M.K. Daly, *Advanced persistent threat*, vol. 4 (Usenix, 2009). (Nov)
6. J. Mirkovic, P. Reiher, A taxonomy of ddos attack and ddos defense mechanisms. ACM SIG-COMM Comput. Commun. Rev. **34**(2), 39–53 (2004)
7. R. Chinchani, A. Iyer, H. Q. Ngo, S. Upadhyaya, Towards a theory of insider threat assessment, in *2005 International Conference on Dependable Systems and Networks (DSN'05)* (IEEE, 2005), pp. 108–117
8. A.N.A. AlFraih, W. Chen, Design of a worm isolation and unknown worm monitoring system based on honeypot, in *International Conference on Logistics Engineering, Management and Computer Science (LEMCS 2014)* (Atlantis Press, 2014)
9. S. Paul, B.K. Mishra, Honeypot-based signature generation for polymorphic worms. Int. J. Secur. Appl. **8**(6), 101–114 (2014)
10. M.M. Mohammed, E. Aleisa, N. Ventura, Zero-day polymorphic worms detection using ahocorasick algorithm
11. P. Jain, A. Sardana, Defending against internet worms using honeyfarm, in *Proceedings of the CUBE International Information Technology Conference* (ACM, 2012), pp. 795–800
12. L. Vokorokos, P. Fanfara, J. Radusovsky, P. Poor, Sophisticated honeypot mechanism-the autonomous hybrid solution for enhancing computer system security, in *2013 IEEE 11th International Symposium on Applied Machine Intelligence and Informatics (SAMI)* (IEEE, 2013), pp. 41–46
13. K. Chawda, A.D. Patel Dynamic & hybrid honeypot model for scalable network monitoring, in *2014 International Conference on Information Communication and Embedded Systems (ICICES)* (IEEE, 2014), pp. 1–5
14. I. Alberdi, E. Alata, V. Nicomette, P. Owezarski, M. Kaâniche, Shark: Spy honeypot with advanced redirection kit, in *IEEE Workshop on Monitoring, Attack Detection and Mitigation (MonAM07)* (2007), pp. 47–52. (ps approach for preventing, detecting, and responding to ddos attacks. Br. J. Appl. Sci. Technol. **5**(5), 500 (2015))
15. R. Selvaraj, V.M. Kuthadi, T. Marwala, An effective odaids-hps approach for preventing, detecting, and responding to ddos attacks. Br. J. Appl. Sci. Technol. **5**(5), 500 (2015)
16. S.S. Sadamate, V. Nandedkar, *Advance honeypot mechanism-the hybrid solution for enhancing computer system security with DoS*, vol. 4 (2015)
17. B.-X. Jia, S.-X. Xie, Dynamic forensics model based on ontology and context information. Netinfo Secur. **1**, 026 (2012)
18. T.H. Project, www.honeynet.org
19. O. Hayatle, A. Youssef, H. Otrok, Dempster-shafer evidence combining for (anti)-honeypot technologies. Inf. Secur. J. Glob. Perspect. **21**(6), 306–316 (2012)
20. S. Mukkamala, K. Yendrapalli, R. Basnet, M. Shankarapani, A. Sung, Detection of virtual environments and low interaction honeypots, in *Information Assurance and Security Workshop, 2007. IAW'07. IEEE SMC* (IEEE, 2007), pp. 92–98
21. X. Fu, W. Yu, D. Cheng, X. Tan, K. Streff, S. Graham, On recognizing virtual honeypots and countermeasures, in *2nd IEEE International Symposium on Dependable, Autonomic and Secure Computing* (IEEE, 2006), pp. 211–218
22. C.C. Zou, R. Cunningham, Honeypot-aware advanced botnet construction and maintenance, in *International Conference on Dependable Systems and Networks, 2006. DSN 2006* (IEEE, 2006), pp. 199–208
23. A. Nicholson, H. Janicke, T. Watson, R. Smith, Rolling the dice-deceptive authentication for attack attribution, in *Reading: Academic Conferences International Limited* (2015), pp. 223–XI, http://ezproxy.deakin.edu.au/login?url=http://search.proquest.com/docview/1781336066?accountid=10445
24. G. O'Gorman, G. McDonald, *Ransomware: a growing menace*, (Symantec Corporation, 2012)
25. C. Seifert, R. Steenson, I. Welch, P. Komisarczuk, B. Endicott-Popovsky, Capture-a behavioral analysis tool for applications and documents. Digit. Investig. **4**(Suppl), 23–30 (2007)

26. C. Sandbox, Automated malware analysis (2013)
27. L. Pearce, Malware analysis in a nutshell. Technical Report (Los Alamos National Laboratory (LANL), 2016)
28. B.M. Bowen, M.B. Salem, A.D. Keromytis, S.J. Stolfo, Monitoring technologies for mitigating insider threats, in *Insider Threats in Cyber Security* (Springer, 2010), pp. 197–217

Chapter 2
Design Honeypots

Abstract According to Lance Spitzner, any information collected by honeypot will be deemed as attack and unauthorized intrusion. Honeypot can be considered in two levels of Taxonomy. The basic level defines the logical order for planning to implement a honeypot. It also explains each category of the honeypot. The advanced taxonomy covers the deeper meaning of honeypot and describes some of the specialized honeypot framework. In this chapter, a roadmap is provided so as to allow the reader to easily grasp the number of homepot frameworks discussed in the Chap. 3 Honeypot developer and researcher faces several challenges such as the type of honeypot to be implemented, types of IDS used and level of difficulty for the hacker. Impropriate decision made may result collecting wrong information or even expose itself.

2.1 The Concept of Honeypot (Basic Taxonomy)

Any movement found in the honeypot is deemed to be malicious and will be treated as an intrusion. The data set collected by honeypot is small and has high value in its content. According to the Lance Spitzner, "A honeypot is an information system resource whose value lies in unauthorized or illicit use of that resource" [1]. It simply means the honeypot is deemed useless if it is not been probed. The unprobed honeypot may reveal other useful information which contradicts to the above definition, they are:

1. The current honeypot technology has exposed itself as a trap
2. The attackers have loss their interest in the current honeypot setup

Figure 2.1 shows the different categories of honeypot in layers in ascending order. Each layer requires decision making before moving down to the next layer.

Honeypots are used for different purposes. They are divided into two main categories, namely for research and production.

The research honeypot is used to collect, monitor and analyze the activities of attacker and the tools used to hack into the honeypot. It is used to discover an unknown vulnerability and attack.

C. K. NG et al., *Honeypot Frameworks and their Applications:*
A New Framework, SpringerBriefs on Cyber Security Systems and Networks,
https://doi.org/10.1007/978-981-10-7739-5_2

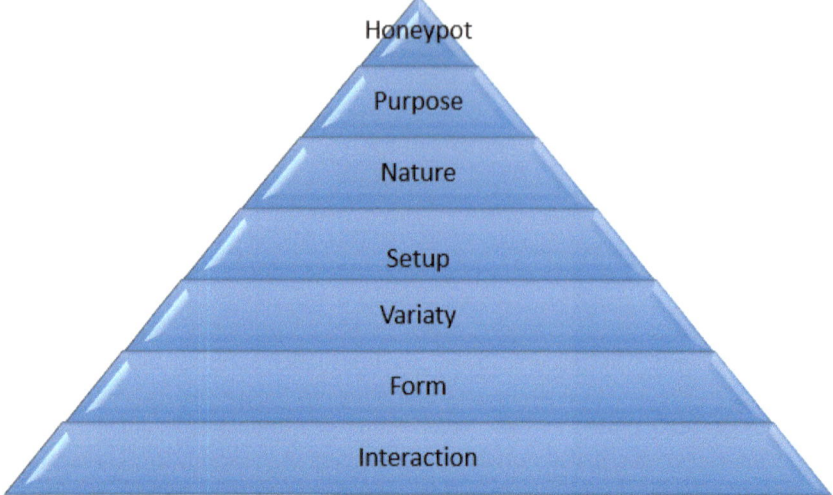

Fig. 2.1 The overview of honeypot concept in layer

Unlike research honeypot, production honeypot focuses in the defensive aspect. It is mainly implemented behind the firewall and conceals within the production network. Its purposes are to keep attacker away from the actual system. It creates an illusion that the attackers are attacking the actual system and alert the system administrator of the intrusion.

Gangfu Feng's proposed linkage defense system design with honeypot [2] is an excellence example of production type honeypot. This proposed system includes honeypot and other traditional security system to protect the production network. There are three important factors contribute to the success of this honeypot framework. They are network data collection, cover and concealment, and logging of all activities. This honeypot framework has the abilities to interact with its owner with expected respond and record all activities of the attacker. It has no difference from the real system in term of respond and time which help to avoid being detected.

The nature of the honeypot is an important factor that needs to be carefully considered. This category can be classified into server-based or client-based honeypot. A server-based honeypot is passive in nature and the client-based honeypot is actively searching for attack over the internet. It is important to take note that all production honeypots to our best knowledge are passive in nature.

Radek Hes proposes an active client-based honeypot framework, Capture-HPC. It actively searches for malicious servers which attack web browser or other application like acrobat reader [3]. The proposed framework consists of a central control server and multiple client honeypot. The server sends a series of command to the client which will in turn act on the command. The task of the client honeypot is to monitor and capture any changes in file system, registry and communication while visiting the targeted web site or remote material. The changes will be communicated back to the server.

The honeypot can be setup as just a simple standalone honeypot, like the framework used in insider detection honeypot [4] or a network of honeypots with a more complex setup, such as multiple worm detection honeynet framework [5]. This complex setup is usually refers to a honeynet or a honeyfarm. The honeynet implementation can also be further classified according to its variety. It can be implemented as a network of single type of interactive honeypot or comprising of severe different interactive honeypots.

Honeypots can also be differentiated by their form. There are two different form namely the virtual honeypot and physical honeypot. Physical honeypots use dedicated host for each honeypot. It is more costly when compare to honeypot which is virtually setup. Virtual honeypot is setup on a single computer with shared resource [6]. These honeypots can be used to imitate a real production network with real working servers and other resources that can produce the same result as of the physically implemented honeypot. The virtually implemented honeypot can be high-interactive honeypot (real operating system with real vulnerabilities) or low-interactive honeypot (emulate service) or a group of honeypots which is also refer to as honeynet.

Xuxian Jiang proposes an interesting "out of box" concept for the virtual honeypot. The "out of box" is fulfilled by moving the anti-virus security suite or monitoring software out of the virtual machine to the host [7]. The author pointed out some of the blind spot and the danger of being detected by anti-malware software and IDS which are implemented in the honeypot with the "in the box" concept. However, the "out of box" concept is not a flawless concept, the semantic view which the "in the box" approach enjoy is lost as only the memory pages, register and disk block can be seen.

This issue can be resolved by using a third party application namely Vmwatcher to provide a non-intrusive virtual machine introspection that will not disturb the system state of the virtual machine being monitored. The guest view casting application reconstruct the semantic-level view of the VM, which bridges the semantic gap which mentioned earlier. The guest view casting also allows the anti-malware software to perform an equivalent "in the box" scanning for virus and malware without the host being affected.

Honeypots themselves, in principle, is classified into two main different types, low-interactive and high-interactive honeypots [1]. The low-interactive honeypot is typically an emulation software tool used to imitate the network services and host systems. It provides limited interaction with the attacker and is generally use to trap and monitor attacker using known attack [8].

Jungsuk Song has include a low-interactive honeypot into his proposed active cooperate-based honeypot to detect the attack [9]. The proposal consists of two parts, a set of low-interactive honeypot and the control server. The honeypot is further broken down into three components, there are the monitoring system (TAP) which is responsible for the communication between attacker and honeypot, honeypot (Nepenthes) that response to the attacker's request and firewall (FW) which allows the communication of the honeypot and control server. The responsibility of the honeypot is to contain the attack as long as possible and the decision of which port to be opened in the honeypot is made by the control server by sending instruction to the FW and the FW will act accordingly. Another function of the control server is to collect and analyze the data from the honeypot.

High-interactive honeypots use systems with real operating system and service virtually or physically for the attacker to compromise it. It provides a lot of interaction between the system and attacker. It allows researcher to discover new kind of attack.

Both high-interactive and low-interactive honeypot do have certain advantages and disadvantages. The advantage of low-interactive honeypot is easy to setup and configure, and its disadvantage is the limitation detect to known threat. As mentioned above, high-interactive honeypot can discover new threat and malware, but it is more complex to setup. Unlike low-interactive honeypot which the monitoring and logging function are include in the emulation software itself; the monitoring software, event logging software, firewall and IDS need to be carefully planned, considered and configured to prevent attacker to use the honeypot for their advantage.

Figure 2.1 provides a hierarchical flow for honeypot decision making. The choice of which interactive type, form, variety and setup are influenced by their purpose and nature. The decision of honeypots may also be guided by the budget and expert available which are not relevant and will not be discussed in this survey.

2.2 Advanced Taxonomy of Honeypot

Basic taxonomy of honeypot covers a superficial explanation of each category of honeypot. In this section, we will like to expand the nature of honeypots further according to their specialization and framework. They will be renamed as attack type. Table 2.1 gives a full review for the advanced taxonomy of honeypot.

Table 2.1 Overview for advance taxonomy

	Attack type	Specialization	Framework
Honeypot	Client-based	Web-based	
		General porpose	
	Honeytoken	Phishing	
		Insider	
	Server-based	Web-based	
		Worm detection	
		Bot detection	
		APT detection	
		General purpose	Dynamic
			AI
			Shadow

In the attack type, the honeytoken has been included as a class of its own. Honeytoken can be passive or active in nature and this will be elaborated further in the next section.

The attack type is expanded and a new category is added called specialization. The specialization class describes the honeypot application for specified attack and the honeypot technique used. This includes:

1. Web server-based honeypot
2. Web client-based honeypot
3. Worm detection honeypot
4. Bot detection honeypot
5. APT detection honeypot
6. General purpose honeypot

The general purpose of honeypot aims to detect more than one type of attack and it can be further expand based on its framework. Thus, framework class is created. These include:

1. Shadow honeypot
2. Dynamic honeypot
3. Artificial intelligence honeypot

2.3 Roadmap of the Honeypot Concept

Figure 2.2 shows a roadmap of the honeypot and the proposed framework to be discussed. The articles are grouped according to their functionalities. There are ten main groups namely, web-based honeypot, worm detection honeypot, bot detection honeypot, dynamic honeypot, shadow honeypot, honeytoken, advance persistent threat detection honeypot and production honeypot.

2.4 Challenges in Designing Honeypot

Honeypots provide us a rich and relevant information about the intruder and his/ her attack. This can only be fulfilled when the honeypot is set up appropriately with certain properties attracting the attacker. Researchers face enormous challenges in the initial setup. Questions such as:

1. What type of honeypot should be used?
2. Which IDS to be included in the research?
3. What kind of attack to be capture?
4. What vulnerability should the honeypot emit?

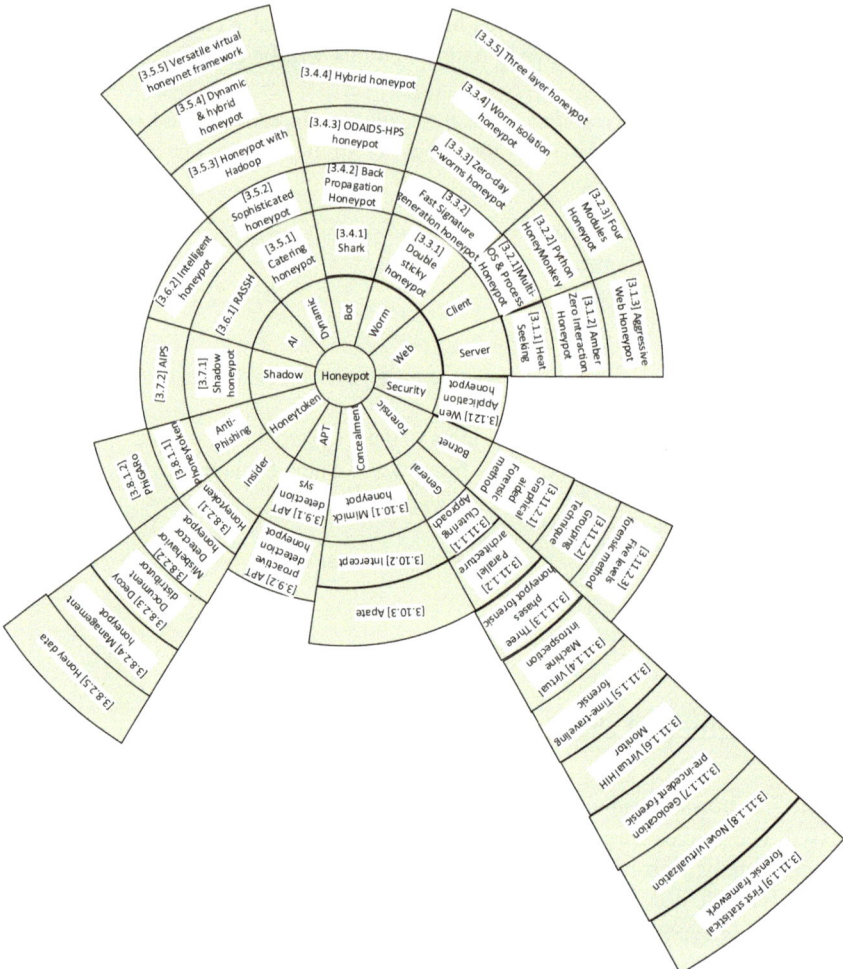

Fig. 2.2 The roadmap of honeypot

Often, researchers are able to capture great amount of attack from the "wild". Those attacks does provide good research value but to a limited extent. This is because most of the attack capture are from script kiddies who take hacking as merely an interest.

Researcher fails to consider serious hacker like the black hat professionals and organized cyber criminal gangs. The use of popular vulnerabilities allow the honeypot to be compromised easily. Thus, the level of challenges to compromise it is relatively low which will fan those people away. What is level of immersion in the honeypot should be? The honeypot owner should also consider making the process of compromission challenging and "interesting" to the expert hacker?

References

1. L. Spitzner, Honeypots: catching the insider threat, in *19th Annual on Computer Security Applications Conference, 2003. Proceedings* (IEEE, 2003), pp. 170–179
2. G. Feng, C. Zhang, Q. Zhang, A design of linkage security defense system based on honeypot, in *Trustworthy Computing and Services* (Springer, 2014), pp. 70–77
3. R. Hes, P. Komisarczuk, R. Steenson, C. Seifert, The capture-hpc client architecture, Technical Report (Victoria University of Wellington, 2009)
4. B.M. Bowen, M.B. Salem, A.D. Keromytis, S.J. Stolfo, Monitoring technologies for mitigating insider threats, in *Insider Threats in Cyber Security* (Springer, 2010), pp. 197–217
5. B.K. Mirsha, U. Kumar, G. Sahoo, in *Double-Sticky-Honeynet for Defending Viruses in Computer Network*, vol. 7 (2012), pp. 131–134
6. L. Spitzner, Dynamic honeypots (2003)
7. X. Jiang, X. Wang, D. Xu, Stealthy malware detection through vmm-based out-of-the-box semantic view reconstruction, in *Proceedings of the 14th ACM conference on Computer and communications security* (ACM, 2007), pp. 128–138
8. D. Dagon, X. Qin, G. Gu, W. Lee, J. Grizzard, J. Levine, H. Owen, Honeystat: local worm detection using honeypots, in *Recent Advances in Intrusion Detection* (Springer, 2004), pp. 39–58
9. J. Song, H. Takakura, Y. Okabe, Cooperation of intelligent honeypots to detect unknown malicious codes, in *WOMBAT Workshop on Information Security Threats Data Collection and Sharing, 2008. WISTDCS'08* (IEEE, 2008), pp. 31–39

Chapter 3
Specialized Honeypot Applications

Abstract In the this chapter, different evolutionary concepts of specialized honeypot to resolve the rapid growing issues of concern of various threats are compared and discussed. As mentioned in the previous chapter, honeypot can be built for a very specified purpose which we will called it 'specialized' honeypot. These honeypot will be used to monitor a very specify malware or attack like worm and advanced persistent attack.

3.1 Web-Server Based Honeypot

Web-server based honeypot is a honeypot which is used to act as a web server. This server, an emulated service or actual server physically or virtually, use to probe the attack. Honeypot often refers to server-based honeypot. The web-server based honeypot waits for attacker to discover its vulnerability and to compromise it. The main purpose of this honeypot is to attract attacks so as to collect information about the malicious activities. Table 3.1 reveals a summary of web server-based honeypot which will be discussed in detail.

When the first web server was introduced in the late 1970s, it is limited to hypertext markup language which is also called html in short. The purpose of the web server is for browsing and provides no interaction between the user and the server. The technology in software and hardware available during that time do have some contribution to this limitation. Then in the late ninetieth, functions such as search, posting and uploading are made available for internet user to interact with the server through the introduction of common gateway interface (CGI). It is well said in the Chinese idiom, "Water can float a boat, but it can also sink a boat", suggests that CGI can be used for good cause and it can be used to conflict damage to the server. Flooding the server through the use of CGI becomes the first web application attack known [4].

New frameworks such as PHP, ASP.NET, AJAX and so on are used to replace CGI, as the technology continue to advances rapidly from the early twentieth-first century up to this present day. The new frameworks provides more interactive features which allow users more flexibility and power to manage data within the web application.

© The Author(s) 2018   15
C. K. NG et al., *Honeypot Frameworks and their Applications:*
A New Framework, SpringerBriefs on Cyber Security Systems and Networks,
https://doi.org/10.1007/978-981-10-7739-5_3

Table 3.1 Summary of the web server-side honeypot

S/No	Ref No	Author	Year	Framework	Type of honeypot	Purpose of honeypot	Form of honeypot	Flexibility	Detection
3.1.1	[1]	John P. John	2011	Heat-seeking Honeypot	Low	Production	Physical	High	Filter
3.1.2	[2]	Adam Schoeman	2013	Amber-Zero interaction honeypot	High	Production	Physical	NA	IP filter
3.1.3	[3]	Supeno Djanali	2014	Aggressive Web honeypot	Low	Research	Physical	NA	NA

Table 3.2 Number of attacks initiated in 2011 by countries [5]

Remote file inclusion		SQL injection		Directory traversal	
Country	Attacks	Country	Attacks	Country	Attacks
USA	20918	USA	91606	USA	189474
United Kingdom	1897	China	47800	Sweden	13535
Netherlands	1879	Sweden	8789	France	9417
France	1253	Indonesia	3604	Netherlands	8320
Republic of Korea	1070	United Kingdom	3419	Germany	7656
Germany	1030	Netherlands	2793	United Kingdom	6692
Sweden	1012	Ukraine	2489	European Union	4159
Brazil	506	Republic of Korea	2374	Canada	3492
Russian Federation	490	Romania	2136	Republic of Korea	2838
European Union	460	Germany	1263	China	2507

The great flexibilities exposes various system vulnerabilities and loopholes. The server becomes vulnerable to attack such as XSS attack, RFI, SQL injection and DT.

Even though patch and new version of the frameworks have released to patch the vulnerable, the blackhat community is still able to find and exploit new vulnerabilities in the server. Honeypot developer has put in tremendous amount of effort to detect and study such attack in order to derive new solution to the problem. Honeypot such as honeyd and high interaction honeypot analysis toolkit (HIHAT) are introduced in the early twentieth century to capture and monitor these attacks.

Table 3.2 reveals number of attack initiated in descending order by countries. The figure is alarming and requires immediate attention.

John P. John in his article has proposed a heat seeking passive web-based honeypot to capture attacker who targets compromised web server with the common server attack. The proposed framework emulates the most popular vulnerability and construct the page which is most probable be attacked by the attacker [1].

The web server-based honeypot is comprised of four components. First, it has a module that identifies the web pages that are targeted by attacker and automatically generate the web page. Second, the web pages are query based and there are no software involved. Third, link is advertised on the search page and all interaction between server and attacker are logged. Lastly, filter is used to separate the attacker traffic and the legitimate user traffic.

This honeypot is easy to implement as it utilizes the advantage of low-interactive honeypot. There are examples to emulate the vulnerability of the software and to interact with the attacker. The low-interaction honeypot implemented gives great flexibility to add new high-interactive honeypot to provide a more realistic environment for the attacker.

Web server-honeypot not only able to provides a decoy environment for the attacker, it helps to consolidate useful information for the filter and firewall to provide more efficient detection of malicious traffic.

Researcher such as *Adam Schoeman* proposes a zero-interaction honeypot-web server to create a blacklist of source IP address. This honeypot combined the best of the two concepts namely decision through detect (DtD) and decision through presence which are adopted by the traditional security system and honeypot for its discovery phase and action phase [2]. The honeypot is assigned to the unused IP address within the web server domain.

In discovery phase, the honeypot spawn a set of common TCP/IP ports on the listener interface and record the source IP address. The record is then compared with the packet collected using a packet sniffing application such as tcpdump. The upstream network enforcer labelled the packet as malicious by checking if the source IP address attempted to connect to other host within the same domain. This web server based honeypot allows itself to be attracted and attacked through DDoS as it is non-productive in nature.

The proposed frameworks in [1, 2] emit the popular web vulnerability to capture the common attacks on server, it does not give reference of which attack it is capable or incapable of capture. In Table 3.2, XSS attack is omitted due to the fact of the difficulty to determine its attack source of origin.

Supeno Djanali has proposed a low-interaction web honeypot equipped with obfuscated javascript code to detect cross-site scripting and SQL injection. The proposed framework is able to complement the limitation in the existing honeypot such as Glastopf, a low-interactive honeypot and HIHAT, a high-interactive honeypot [3]. The web honeypot, consists of three different web pages, is implemented using the likejacking technique and the hardcoded script of the commonly used SQL injection attack.

The web honeypot main page portraits an institutional web site to lust the attacker preform XSS or SQL injection to compromise the web site. The obfuscated javascript analyzes and redirected the request to the appropriate page (XSS page or SQL injecting page) according to the nature of the request. The requested page will emulate the vulnerability anticipated and respond to the attacker. It also records the information of the attacker such as identity, browser, agent information and operating system fingerprint.

Summary for web server-based honeypot: Table 3.3 reveals the number of attacker IP detected visiting the honeypot. This result exclude the number of repeated IP occurrence. The variation in the figure is affected by several factors such as:

Table 3.3 IP address capture (No repeat)

S/No	Framework	IP address detection
3.1.1	Heat-seeking honeypot	6438
3.1.2	Amber-Zero interaction honeypot	529
3.1.3	Aggressive web honeypot	610

Table 3.4 Average visit for each honeypot

S/No	Framework	Total visit	Average visit
3.1.1	Heat-seeking honeypot	44696	7.5
3.1.2	Amber-Zero interaction honeypot	4132	6.98
3.1.3	Aggressive web honeypot	36000	59

1. The purpose of the honeypot
2. Location of the honeypot

It is obvious that the total number of visit capture is not equivalent to the number of IP address and the average number shows in Table 3.4 reveal this fact. Table 3.4 shows sign of multiple visit to the honeypot from each IP capture in Table 3.4.

3.2 Web Client-Based Honeypot

Client honeypot is a honeypot that actively search for malicious or compromised server which attack on client. It acts like a normal computer or application and interact with the server to examine whether an attack has occurred. The main objective of the client honeypot is to capture and identify the malicious server by actively visiting them (Table 3.5).

Basically, there are two types of attacks namely server-side attack (which has been described in detail) and client-side attack. Contrary to server-side attack, client-side attacks are those attacks focusing on the vulnerabilities of client application such as web browser, email and office software.

In recent years, attack on client side vulnerability has been increasing and there is a need of new tools and techniques to defend such attack. Tools such as IDS, firewall and anti-virus software are based on the pre-defined signature available within the application database to detect attack and prevent the network from known attacks. These security tools do have their drawback such as inability to detect zero-day attacks that there are no signature within the database.

Honeypot plays a big role by tighten the network from these kinds of attack. It is worth to mention that client honeypot does not provide direct security or protection to the client host or network. It gives us useful information so that we can take further

Table 3.5 Summary of Web Client-side Honeypot

S/No	Ref No	Author	Year	Framework	Interaction of honeypot	Purpose for honeypot	Form of honeypot	Detection method
3.2.1	[6]	Mitsuaki	2012	Multi-OS and Multi-process honeypot	High-interactive	Production	Virtual	Hooked API function on file operation file finding, registry operation, process creation, process termination and code injection
3.2.2	[7]	Rohit Shukla	2014	Python honeymon-key	High-interactive	Production	Virtual	Signature-base IDS
3.2.3	[8]	Tung-Ming Koo	2013	Four modules client-side honeypot	High-interactive	Production	Physical	Source code analysis module behavior analysis module

remedial action and also have a deep understanding of client-side attack. Generally, client honeypot is made up of three components, queuer, client and analysis engine.

The queuer consolidates all malicious web site and create a list of server for the client to visit. The initial consolidation process plays an important roles here. It is the eye that pin point the client to the correct malicious web site. It will cause the honeypot to be deemed useless if it does not perform its task correctly. *Rohit Shukla* proposed malicious internet web site collection system which is similar to honeymonkey introduce by Microsoft [9]. The proposed system is used to collect and analyze the information of the malicious web site. It comprised of four virtual client high-interactive honeypots of various operating system [7]. The master system is setup in isolation to other network rather than the one connected to the client. The master system control the slave/client system and it also hosts database log server. The honeypot consists of a snort IDS running at the background and web clawer.

The web crawler which is the focus, act as a queuer, takes in the web parameter, extracts all the links and URLs, and stores in the database in the master system. Based on the information collected, the master system issues command via the secure shell process for the honeypot browser to visit the URL link. The IDS in the honeypot record and store the data for event occur and IP address which match the signature into an external database. A list of blacklisted web site is compiled into a file.

The client visits the web site and contain the attack. The client is the actual honeypot that collect important information about the attacker. It is required to collect as much information as possible about the attack. Any misinformation can affect the level of accuracy in the next component. *Mitsuaki* has proposed a client honeypot system that uses multi-operating systems and multi-process honeypot multiplication approaches using the web browser to provide high scalability and performance-efficiency. The purpose of this proposed honeypot architecture is to detect drive-by download from the malicious web site. Multiple honeypot instances are created in a single physical machine or virtual machine [6]. The increasing instances of honeypots require many OS instance in a physical machine or virtual machine to improve the inspection performance. According to the article, there are two processes proposed, namely multi-OS and multiple processes, which make the high interaction honeypot system to be scalable and efficient.

The multiple-operating system process is made up of three components. They are honeypot-agent which is a parent program of the web browser on the honeypot instance, honeypot-manager a controller used to provide instruction to the agent via agent processes and, virtual machine monitor that use for each honeypot operation and also provides virtual space for each OS. Thus, this allows multiple OS to be ran on a single physical system. The multiple process concept in the web-based honeypot allows the OS to launch other browser processes when the current browser is idle. This reduces the OS overhead and improve the inspection efficiency. Process isolation mechanism such as process sandbox is implemented in the process multiplication solve the issue of not able to determine the cause of exploitation in the event of anomaly occurred.

The analysis engine analyses and determines whether an attack has occurred on the client honeypot. For better detection of the latest client-side malware, researcher [8] has focused on creating a better analysis engine of the client honeypot. *Tung-Ming Koo* proposed to use client honeypot with custom static and dynamic analysis engine to detect malicious website that download malware stealthy via drive-by method into the client [8]. The system consists of four modules. They are the proxy module, source code analysis module, behavior recording module and behavior analysis module. The proxy module record the address, save the web page and send it to the source code analysis module. It then waits for the result before decide whether to allow the web page to be sent to the client.

The other three components made up the analysis engine. The source code analysis module based on the static content analysis to test the web under those assumptions. The first is the obfuscation coding in the webpage script, application attack or leakage attack by redirect user to malicious image files and second is the present of abnormal semantics in the source code. After testing using the above mentioned assumption and the web page is yet to be deemed to be safe, will be passed to the next module record behavior module. This module is operated on the captureBAT, a client honeypot. In this module, the web page is executed on a simulated windows environment to capture the event such as changes done to the registry, I/O and the construction and destruction of processes. The captured information is passed to the last module for analysis. The last module analyzes each event and identifies the severity level of damage. The level of damage is determined by analyze three major events namely I/O event, registry event and process event. The web page will be considered malicious if the signature of any the event response negatively. Thus, the user will be prevented to open the web page.

Difference measuring method has been used to determine the level of effectiveness for different proposed honeypot framework. As each author uses the dataset that flavor his/her architecture, the result of each framework is extracted and formulated in

Table 3.6 Result of IP capture using Client-based Honeypot in percentage

S/No	Framework	Malicious detection		Benign detection	
		Success (%)	Fail	Success	Fail
3.2.1	Multi-OS and Multi-process honeypot	100	0	100%	0
3.2.2	Python honeymonkey	100	0	0	0
3.2.3	Four modules Client-side honeypot	100	0	100%	0

Table 3.6. The ability to detect malicious web site is selected as a form of measurement to determine the efficiency of the framework.

Summary for web client honeypot: Although each framework using their own dataset shows an above average result, this result does not give a full picture to justify the effectiveness of the honeypot. This is due to the different in number of dataset used. Example framework two uses only a web with hundred link which twenty-seven of them is deemed to be malicious. If the data-sets which are used in the other two frameworks is considered in former framework, the outcome might only render not more than 85% effectiveness.

In Table 3.6, framework in [6] uses the hooked API to monitor the changes take place in the honeypot, whether the attack is known or unknown is no longer relevant. As all attack is detected and monitored once modification occurs. The framework in [7] is limited to detect known attack which is made up a major portion of the attack. It does not have the ability to detect on unknown attack. Framework in [8] uses the static detection method to detect the known attack and dynamic detection method to detect the unknown attack. It basically provides detection for all attacks.

The framework in [6, 8] are equally effective when detecting unknown attack. The only issue is that both framework rely on the ability to detect anomaly signal from the traffic which do require time consuming learning process.

3.3 Worm Detection Honeypot

Worm attack has and will always be one of the top focus by the system administrator and security software developer. Over the years, worm has done some of the most devastating damage that caused the loss of billions of dollar economically. The worm detection honeypot helps researcher to study the worm and also to derive a better solution to solve the new strand worm virus.

The term "worm" has gained its notorious reputation over the pass decade for its ability to self-replicate and spread without aid from any form of applications. Worm was first appeared in a 1975 novel called Shockwave Rider where the leading actor designs and sets off worm in the act of revenge against the powerful man who own an electronic information web site. The initial intention and use of worm were for

Table 3.7 Summary of polymorphic worm detection honeypot

S/No	Ref No	Author	Year	Framework	Multiple honeypot	Type of honeypot	Form of honeypot	Detector	Signature generating concept
3.3.1	[10]	Upendra Kumar	2012	Double sticky honeynet	Yes	High	Physical	Gate translator	Position-aware distribution signature algorithm
3.3.2	[11]	Sounak Paul	2014	Honeypot based on fast signature generation system	Yes	Low	Physical	Signature-based detection and anomaly-based detection	Probability calculation of the multiple invariant strings present in a polymorphic worm
3.3.3	[12]	Mohssen	2014	Zero-day polymorphic worms honeypot	Yes	High	Physical	Gate translator string matching algorithm	Aho-Corasick Algorithm
3.3.4	[13]	AlFraih Abdul	2014	Worm isolation honeypot	Yes	High	Virtual	Anomaly-based detection	Auto signature generation
3.3.5	[14]	Pragya Jainc	2012	Three layer honeypot	Yes	High and Low	Physical	Signature-based detection and anomaly-based detection	Netbeans 7.0

good cause such as network worm created by John Shoch in Xerox PARC in 1982 to monitor the performance of the ethernet principle in the network.

Unfortunately, most of the later development of worm move to join the dark side such as the first worm by Robert Morris which infected a tenth of the computer online and the latest worm such as NGRBot worm which help bot master to build up the botnet using the IRC channel. Worm can be used to convey message from the social hatred to the CEO of mega software company such as message to Bill Gate, CEO of Microsoft Enterprise, embedded in the string of Blaster worm serve as a wake up call.

The use of worms have gave security researcher a hard time looking for alternative way to detect and study them. Current security system such as anti-virus software and firewall provides good protection to keep the worm away. It does not provide a 'space' for the worm to mingle with the system so that researcher can do more studies to understand them. Honeypot answers this call. Multiple honeypots [11–15], which can also refer to honeynet, provide a playground for the worm to mingle with severe hosts. All activities occur in the honeypot and within its network are logged.

Upendra Kumar in his articles has proposed a framework with multiple honeypot concepts to capture the polymorphic worm [10]. The framework consists of three honeypots, gate translator and router. The gate translator separates and redirects the infectious packet to the inbound honeypot. The honeypot with the signature-based intrusion detection algorithm detect the old virus and its packets. The outbound honeypot with the Position-Aware Distribution Signature algorithm verifies, records and analyses the new evolved worm virus. Lastly, the worm is transferred to the sticky-honeypot. The activities of the worm virus halt once it is being redirected into the unused IP address system which is equipped with the updated anti-virus software.

Upendra Kumar, later, introduces a similar framework as the first concept in his second article, the major difference is that it is using two honeynets to capture the worm [15]. Both concepts provide room for the worm virus to manoeuvre and evolve within the space specified. The second concept has the abilities to generate more signature of the polymorphic by allowing the worm to roam within the honeynet.

Sounak Paul proposes the use of multiple low-interactive honeypots to capture polymorphic worms. The proposed framework is less complex in setting up than [15]. It adopts the fast signature generation scheme which is based on the probabilistic approach considering its abilities to generate an accurate polymorphic worm signature even noise exist [11].

The framework makes room for the worm to roam between the two sets of low-interactive honeypots which emulate different types of commonly used services such as DHCP, FTP, HTTP and POP3. The information of the worm is collected using the multi-layer data capturing system. The first layer is the entry layer where the firewall logs the packet header once the worm enters the network. The second layer is the layer between the router and the honeypot sensor, full packets are capture and saved. The last layer is the honeypot which records the activities of the attacker.

A restrictive path method can be applied in the honeypot framework to limit the movement of the worm so that the worm will not move beyond the boundary. This allows the researcher to manipulate the worm towards his/her desire direction and an extra precautious step to reduce the probability of out of control. *Mohssen* proposes to capture of the variant of polymorphic worm by allowing it to interact with two set of honeypot in the planned fashion. The major distinctions between [11, 12, 15] are the algorithm used for signature generating and the degree of roaming freedom for the worm.

In this honeypot structure, the worm was given limited freewill to roam freely within the honeypots [12]. Every movement of the worm is directed by the internal translator one and two which act as the door for the honeypots. The worm is redirected from honeypot group one to group two by the internal translator one and via visa. The signature generator adopts a dictionary matching algorithm to match the worm instant and accurately create the signature for the worm.

Honeypot, in all instant, provides an environment to monitor the worm evolution and it relies on the intruder prevention system and intruder prevention system to detect and notify its user about the intrusion. One of the great skill of polymorphic worm is its ability to replicate and create another variant of itself through the process like obfuscation etc. Signature-based IDS may not be able to detect all of its variants. Anomaly-based IDS can be used to resolve such problem.

AlFraih Abdul proposes a system that combine idea used in sweetbait [16] and Honeystat [17] to setup a production honeypot against polymorphic worm [13]. The burden of detecting the new bleed of worm relay on the network gateway. The gateway adopts anomaly detection method to detect the new unfamiliar data flow by comparing the data against the whitelist. The traffic is then redirect towards the honeypot group. The honeypot group is made up of several virtual honeystat honeypots which collect memory, disk write and network event [17] and store them in the security management center [16]. The honeypot is also equipped with the automated worm signature generation to create the new signature and store them in the security center. The security center updates the signature to the NIDS and NIPS once it is created.

Single IDS reduces the chance of worm detection. For example, using merely signature-based IDS can detect the known worm but not the unknown. Both types of IDS can be used together to ensure the known and unknown worm do not slip through the check.

Pragya Jainc proposes the use of three layers concept with the honeyfarm to help to detect new unknown worm with polymorphic, monomorphic and metamorphic in nature [14]. The first layer of defend consists of signature-based intrusion detection system to detect the known danger based on the signature comparison. The second layer used anomaly-based intrusion detection system to detect the unknown worm by detecting the abnormal behavior. The last layer is made up of a group of high and low interactive honeypot. The proposed system uses a roaming method to randomly select a honeypot to play the role of control center for security purpose.

The proposed system filters the known threat and redirect the unknown threat to the honeyfarm so that the activity can be analyze and new signature can be created accurately to reduce the false positive in the signature-based IDS.

Summary for worm detection honeypot: The initial detection of polymorphic worm plays a very vital role in worm detection honeypot. According to the Table 3.7, signature-based IDS is commonly include either as the main IDS or part of the IDS system. The framework proposed by [10] configures the edge router to recognize the list of unused port in the process of data accession. The framework proposed by [11, 14] use two different type of IDSs to detect the worm attack. Reference [12] uses only signature-based IDS.

Table 3.7 shows framework proposed by Upendra [10] and Mohssen [12] are useful for generating the variant of the known polymorphic worm. They are limited in capacity to deal with the known polymorphic worm due to the nature of IDS used. The focus is to collect as much variant as possible for the known polymorphic worm. The framework proposed by [11, 13, 14] use anomaly-based IDS for the detection which is able to detect the unknown and also the known worm. Anomaly-based detection does have disadvantage, it can result high false positive if insufficient training is provided for the IDS or the IDS is not tunes to the appropriate level.

In all frameworks, signature generator is adopted to create the signature for the polymorphic worm. References [10, 13, 14] focus on how the worm is trapped in the honeypot and collect as much as possible of the variant of the worm. References [11, 12] frameworks have given a considerable amount of detail on their proposed frameworks, algorithm for extracting the variant and test result base on the dataset used.

Reference [12] does mentioned about how the worm can be capture and explain how Aho-Corasick Algorithm can be used to generate the obfuscation. This algorithm uses the string searching method to locate a finite set of string in the dictionary to match all patterns. The [11] treats all string as token and a token can be presented in a significant number of flow. Signature is a set of tokens with their occurrence number in suspicious flows.

To conclude, the proposed architecture in [10–14] adopt a fairly similar idea with multiple honeypots to capture the signature of the worm. The ability to capture all the variant using honeypot or generate signatures to accurately represent all the variant of the polymorphic worm is vital. The first, forth and fifth framework use generator that is limited to generate variant only collected from the honeypot. Comparing the algorithms in [11, 12], there seem to have a gap in the level of algorithm advancement even though both articles are published very recently. The Aho-Corasick algorithm which is implemented in the architecture of [12] stand out. The Aho-Corasick Algorithm is able to generates signature even if not all the variant of the polymorphic worm is capture.

3.4 Bot Detection Honeypot

Bot is also called zombie host and botnet is a network of zombie hosts. Botnet can be used for many malicious acts such as sending perishing mail, launching Distributed Denial of Service attack and collecting user information. Among all, DDoS attack has the most devastating effect. The DDoS attack can be used to paralyze the set of servers or single server to affect their abilities to perform the normal daily tasks.

Table 3.8 Summary of the DDoS dectection honeypot

S/No	Ref No	Author	Year	Framework	Type of hon-eypot	Form of honeypot	Technique
3.4.1	[18]	Ion Alberdi	2007	Shark	High	Physical	Redirection kit
3.4.2	[19]	Sherif Khattabv	2006	Back propagation honeypot	High	Virtual	Server replica and back propagation
3.4.3	[20]	Rajalakshmi Selvaraj	2015	ODAIDS-HPS honeypot	High	Physical	Outlier IDS
3.4.4	[21]	Swapnali Sundar Sadamate	2015	Hybrid honeypot	Low	Virtual	Threshold-based anomaly IDS system

Botnet is very hard to detect by a single border firewall. The implementation of Bot detection honeypot can be for two purposes. It can be used to contain bot and study its origin or can be used to contain the DDoS attack in order to keep the real server safe and free from such attack (Table 3.8).

The concept of bot was initially used to serve the community; the benign bot, eggdrop bot, used for automating the basic task on Internet Relay Chat (IRC) [22]. In the late ninetieth, the first two malicious bot, Sub7 or Pretty Park, which were actually a trojan and worm, are created to connect to the IRC channel and wait to receive command from the bot master.

In the early part of the millennium, the bot writer has adopted difference types of botnet such as http botnet and peer-to-peer botnet which made security researchers work overtime to seek for solution to tackle this issue.

Botnet can be detected by honeypot with intruder detection system. The honeypot allows the bot master to compromise so that any movements and malicious activities can be monitored and analyzed. The IDS either adopts static analysis (signature-based) or behaviour analysis (anomaly-based) [20, 21] to detect botnet. Static analysis method involves checking the traffic against a list of known malicious whereas behaviour analysis method monitors the communication of the network for behaviour exhibited by botnet. Behaviour analysis method allows high degree of flexibility while adopting various methods and have great potential in the area of research.

Swapnali Sundar Sadamate has proposed a hybrid system to detect DDoS. The hybrid system adopts the client-server architecture. The system consists of a server that store all information collected by the honeypot, a honeypot to capture the attack information, web-management interface and a threshold-based anomaly IDS system.

The IDS system is implemented at the gateway of the network to detect the malicious traffic. The malicious packet is redirected to the honeypot. The honeypot is install with sebek to record attacker behavior, dionaea to collect the malware information and snort in the verification process to collect and analyse the packet received. The server receives all the data from the honeypot and stores them in a database. The web-management interface allows the data to be presented visually.

The proposed system use the threshold-based anomaly IDS to detect and identify the attack by comparing the incoming traffic with the legitimate traffic records [21]. The result in the differences is examined against the predetermined threshold to determine whether the packet is a DDoS attack or not.

Rajalakshmi Selvaraj implemented an outlier IDS in the honeypot framework to detect the DoS packet. The IDS adopts distance of the nearest neighbor method and requires a set of pure normal data to train the system [20]. The system consists of outlier IDS, attack classifier and honeypot.

The packet is checked and valued by the outlier based IDS. The IDS also outlines the feature of the packet such as percentage of connection having the same destination and same service and percentage of packet with error. The attack classifier will compare the packet outlier value with the threshold to determine whether the packet is DoS packet. The malicious packet is redirected to the honeypot. The honeypot will respond to the packet with the relevant error message to the sender.

Unlike other security method, the activity, such as outgoing traffic, of honeypot is restricted by law in most countries. This makes honeypot a less favour method in the area of research. However, this does not prevent [18] to propose honeypot framework to detect bot.

Ion Alberdi has proposed a system that first use passive network monitoring techniques for observing and analyzing attacks and the spread of bot via malware. The purpose was to discover the behavior of malware which drive the bot [18]. The spreading of the bot is then halted by redirecting the packet to another honeypot within the network using the advanced redirection kit.

The redirection functionality creates an illusion to the attacker that they are able to connect to the internet and the bot is communicating with each other within the botnet. This will also to deduce the possibility of the honeypot to be exposed.

Honeypot can also be used to prevent the actual server from falling victim under DDoS attack, especially, in the commercial environment such as stock market where millions of dollar may be lost even the server is down for a few minutes.

Sherif Khattab has proposed the use of roaming honeypot with back-propagation ability to prevent non-spoofed service-level DoS attack [19]. The honeypot is hidden within the pool of server replicas. The set of server replicas will be selected to become active for a duration of time while the remaining idle servers will act as honeypot. The active server replicas will coordinate with legitimate user. This makes it very tedious to identify the real server and thus trap DoS attack in a honeypot. The proposed framework is based on the assumption that even if the attacker knows all the server and honeypot at time t, he will still not be able to differentiate the honeypot from the real server at time t+1 [23]. The honeypot also have the ability to drop all attacks once it changes from idle to active state.

The back-propagation function of the honeypot allows the server to send out a recursive trace back process by alerting the Autonomous System across the path

Table 3.9 Task of the DDoS honeypot

S/No	Framework	Technique	Detect DDoS	Capture DDoS	Protect from DDoS
3.4.1	Shark	Redirection kit	Yes	Yes	No
3.4.2	Back propagation honeypot	Server replica and back propagation	Yes	No	Yes
3.4.3	ODAIDS-HPS honeypot	Outlier IDS	Yes	No	Yes
3.4.4	Hybrid honeypot	Threshold-based anomaly IDS system	Yes	Yes	Yes

towards the bot. The alert triggered the AS-level input debugging process traffic that are for the honeypot. Access routers of attack hosts or bots are identified and filtering rules are installed to drop all traffic destined to the honeypot. The back-propagation function also help the router to distinguish the attack packet from the legitimate packet so that the attack packet will be dropped to prevent the network to be over-congested.

Summary for bot detection honeypot: As shown in the Table 3.9, most of the proposed frameworks, except [18], focus on detecting DDoS attack on the server. Reference [18] uses a very primitive approach to detect bot by detect and redirect its out-going packet. Reference [19] uses roaming method so that the attacker will not be able to distinguish which is the real server while performing a DDoS attack. Lastly, [20, 21] uses an anomaly-based IDS to identify the DDoS attack and redirect them to a honeypot.

However, all proposed methods do post unsolved question. For [18], question likes what is the probability that the attacker will fall into such illusion that the bot (honeypot) is communicating with each other and will not suspect that he/she is communicating with a honeypot should be carefully considered. For [19], it is just like throwing a dice, the attacker have 50% chance to accurately get the server if he does a DDoS attack at random. The detection of the DDoS attack occurs only after the event happens. For [20], the nearest neighbour method for the IDS do have downside such as high resource consumption and long processing time. Reference [21] seems to have an excellence setup for bot detection, only under one condition. That is no outgoing traffic restriction imposed on the honeypot. The intend setup of the honeypot seems to be redundant if it is only used to contain DDoS attack. DDoS attack is the one of the final products create by botnet and it has little value for research purpose. The initial exploitation is the real juice for the research. It provides information such as method, malware used in the exploitation and also the new vulnerability in the existing system.

3.5 Honeytoken

The concept of honeytoken is as old as the security itself. Honeytoken has the same properties of that of the honepot except that it is not a computer [24]. A honeytoken is digital entity perceive to be valuable by the ignorance attacker. It can be as simple as an ID with the password and also as complex as the spreadsheet with customer information. Honeytoken does, however, face several difficulties in creating especially to generate those spreadsheet with fake customer details. Before generating the honeytoken, there are three main questions that should be answered. They are:

1. How it should be constructed?
2. Who is the honeytoken for?
3. What information should be changed or unchanged?

The knowledge to generate the honeytoken is vital and the honeytoken may be manually generated. The whole process is tedious and time consuming.

Maya Bercovitch et al. has solved the process issue by introducing honeygen, an automated honeytoken generation software that automatically create the complex honeytoken. The honeytoken generation adopts the constrict satisfaction problem approach to generate a honeytoken [25].

The application has two difference modes which can be used to create the honeytoken. The first mode is the obfuscation mode where the real data is used as input. This mode only changes the more sensitive value. The second mode is the generation mode. This mode creates the honeytoken from scratch based on the given rules. The rules is a set of predefined attributes by its user. The information in this mode of honeytoken is artificial and the amount of records is rely on the definition of its user.

The advantage of honeytoken can be best demonstrated in the area of anti-phishing and insider threat which will be discussed.

3.5.1 Anti-phishing Honeypot

Phishing is an attempt to acquire personal credential, often for malicious reason, by pretending to be trustworthy entity such as local authority or staff from financial institution via electronic communication. It has being and will be an ongoing issue and concern for security researcher as more advance technique has been used to avoid being detect by the anti-phishing software. Phishing attack has caused loss of millions of dollar economically.

The most common phishing tactic is to send spam email to as many internet users as possible hoping that someone will be convinced by the content and act accordingly. Majority of the victims are internet novice who does not have adequate knowledge of such danger.

Phishing attack has been around since 1995, it is not commonly known by people until ten years later. The first recorded mention of phishing attack takes place in America Online (AOL). Hacker imposes as AOL employee and request users to verify their account or billing information via AIM accounts. Such account cannot be punished by AOL TOS department and eventually force the company to include warming, which is the first security measure against phishing, in its email and messenger (Table 3.10).

The technique for phishing attack has not change much, but its target has shifted from email and communication software to financial institution and online payment system in the early of second millennium. Email worm program is used to send spoofed email to paypal which will direct those customer to a spoofed site for them to update their credit card detail and other sensitive information.

Different types of phishing attack, such as phishing, spear phishing, clone phishing and whaling, have been used to cause damage economically. Some of the attack are group focused such spear phishing which directed at specific organisation and whaling which only focus on higher management personnel in an organisation. Phishing attack like clone focuses on the technique used to create an indistinguishable fake email to lure user to believe its legitimacy.

Large organisation and institution have setup anti-phishing detection software in their mail server to detect spam mail. Such technique is not 100% foolproof as small percentage of the spam mail manage to get pass the check. Staff training do help to prevent phishing as well. This, however, is not enough, phishing technique such as clone phishing can confused user who fail to distinguish fake email.

Honeytoken and honeypot, in this case, can be setup to attact the phisher to steal from it so as to keep the real system secures. *Shubhika Chauhan* has proposed the use of honeytoken to capture the phisher activities. Shubhika also mentioned the issue with the accessibility of the fake credential and, honeypot vs real online system [26]. In order to overcome such issue, the proposed framework involves part of the legitimate banking system and the system administrator of the bank. The framework consists of honeytoken (phoneytoken), honeypot (honeyed) and spamtrap.

The spamtrap is used to detect the spams and phishing email. The honeytoken generates the fake user credential that is accessible to the real online system. The real system with the knowledge of the honeytoken redirects the phisher to the honeypot. The honeypot used is not a real honeypot. It is a topped-up online banking system with additional features where the bank administrator has a fair share of control over it. To avoid being detected by the phisher, he is allowed limited access to some banking feature such as transferring a limited fund and viewing bank statement.

Honeytoken allows phisher to use the fake credential to log into the intended system so that researcher can monitor and track him/her down. Honeypot can be used to contain the phishing email to assess its level of damage to the system. *Martin Husak* has introduced the use of honeypot as part of the automated detection process for spam and phishing email. The author stated the tedious process for manually process each email to detect phishing and also the report from user does not reliably capture all the phishing email in the network [27].

The proposed framework consist of two parts namely phishing detection and phishing incident processing. The phishing detection unit is made up of a high-interactive honeypot mail server with specify filtering rule suite for phishing detection

Table 3.10 Summary of phishing honeypot

S/No	Ref No	Author	Year	Interaction of honeypot	Form of honeypot	Purpose of honeypot	Honeytoken propagation	Email server	Phishing detection	Capture activity
3.5.1.1	[26]	Shubhika chauhan	2014	Low-Interactive	Virtual	Research	Passive	None	Spamtrap	Honeypot
3.5.1.2	[27]	Martin Husak	2014	High-interactive	Physical	Production	Active	Honeypot mail server	Spamtrap	None

Table 3.11 Features for phishing honeypot

S/No	Honeytoken propagation	Email server	Phishing detection	Capture activity
3.5.1.1	Passive	None	Spamtrap	Honeypot
3.5.1.2	Passive and Active	Honeypot mail server	Spamtrap	None

and the spamtrap installed. The email address of the honeypot or honeytoken is made known to the phisher via active and passive propagation. The honeypot accepts all incoming email and does not forward or send message. The report of phishing email is then reported to the incident processing unit. The phishing incident processing is taken care by PhiGARo which automatically handle the phishing incident when any phishing is reported to the system. The phishing incident process first start by determine whether is the phishing material a URL or an email. Then it checks the material with its database. Lastly, it interprets the result. The phishing incident processing unit too accepts report from human. The process includes blocking the malicious web site, update the phishing filter and inform the victim.

Summary for phishing detection honeypot: The two authors are engaging the issue of phishing in a different manner. Table 3.11 shows framework in [26] focuses on capturing and monitoring the activities of the phisher while the framework in [27] is more on how to detect and prevent phishing. The honeytoken in both architectures is used for different purposes. In [26], it is used to propagate the fake credential to the phisher. In [27], it is used to propagate the existent of the mail server honeypot to the phisher so as to receive phishing email. Both architectures have similarity such as using spamtrap to detect the phishing email. The study of phishing does come with a price, example phisher logs into the bank account and does a bank transfer. Reference [26] has included this as the disadvantage of the proposed framework and he also stresses that the bank should absorb all the expense.

The framework in [26] does have other setback which is not with the concept introduced but with the willingness issue. The bank may refuse to cooperate with the researcher to share its technology and the cost which is mentioned earlier on.

The framework in [27] does also have setback in regards to the detection of the phishing email. It relies on two sources to detect phishing email, one is the email received in the spamtrap and the report of phishing email from the legitimate user. There is no mentioned about the detection of the phishing email from the mail server, where the phishing email may slip through the eye of the ignorance user.

3.5.2 Insider Detection Honeypot

Insider attack is described as the damage to the interests of an organization by a trusted individual with legitimate access to its network and system resources.

Insider attack can be differentiated into two categories. One is the masqueraders who pretend to be another system user and the other is traitor who has his/ her own legitimate system credentials [28]. The motive of insider attack can also be classified into two difference types namely, inadvertent (unintentional) and intentional.

Unlike external intrusion, internal intrusions are most likely done by trusted employee or member in the management who may know the system well. Their objective is not to destroy the system but after some information of high value which is not available to the public for personal gain [29]. This act can be more devastating than the damage done from the outside.

The current insider detection system can only detect the threat post-incidentally and by then it will be too late as damage has taken place. Honeypot adopts the pre-incident approach which can be uses to attract insider attack. This helps to ensure the safeguard of the real system. One of the benefits of honeypot is it helps to improve the management function in an organization system by revealing its problems via compromised. *Li Hong-Xia* has proposed a management honeypot which used to collect, detect and record the attack on management system. The management honeypot borrows the idea of the network honeypot to reduce the deficient and patch loophole of the management system.

The honeypot uses the game theory of the management, relevant mathematical model and other management technologies [31]. The author defines the desire of power and distribution of benefit as honey and the management system with loophole as container fill with honey. As long as there are people try to pass through the loophole for the honey, the game exists. The management honeypot is used to serve as an early warning and is effective in capturing intruder internally.

Most researchers [24, 28, 32] propose the use of honeytoken with honeypot to trap the insider by placing the token in an easily accessible location in the system. This can be within the document folder, in an email application or within a honeypot.

Lance Spitzner presumes that the insider possess the knowledge of obtaining a very specific information. He proposed the combined use of honeytoken and the honeypot to detect insider attack. Three different honeytoken setups are proposed that will trigger the alarm once insider is detected [24]. First, the honeytoken will be triggered when the insider acts inappropriately. The honeytoken in this example contains the user ID, password and the location of the fake server. Second, honeytoken can be concealed in the file and email environment to be perceived as information of high value to attract insider. Lastly, the honeytoken can also be implemented as part of the organization search engine. It will be triggered when the insider perform an unauthorized search or input highly sensitive keyword into the search engine.

The purpose of the honeytoken is to provide user ID and password that allow the insider to use them to gain access to the server (honeypot) to retrieve a particular piece of information which is believed to be of high value for personal gain.

Honeytoken can come in a more complex form such as document. A fake document can be used to attract insider. Unlike the above mentioned, this document contains hundreds of fake customer detail which is very tedious and time consuming to create. Honeytoken generator has been introduced to assist security personnel to

Table 3.12 Summary of insider detection honeypot

S/No	Ref No	Author	Year	Framework	Interaction of honeypot	Form of honeypot	Purpose for honeypot	Decoy form	Detection
3.5.2.1	[24]	Lance Spitzner	2003	Honeytoken	High-interactive	Physical	Production	Honeytoken	Misuse honeytoken
3.5.2.2	[30]	Praveen J U	2013	Honeypot with misbehavior detector	High-interactive	Physical	Production	Honeypot	Misbehavior analysis
3.5.2.3	[28]	Brain Bowen	2010	Decoy document distributor	High-interactive	Physical	Production	Honeytoken	Misuse honeytoken
3.5.2.4	[31]	Li Hong-Xia	2010	Management honeypot	High-interactive	Physical	Production	Honeypot	Attack on loophole of management system
3.5.2.5	[32]	Huseyin Ulusoy	2015	Honey data	High-interactive	Virtual	Production	Honeytoken	Misuse honeytoken

create such document. Before creating this document, several questions need to be considered:

1. Who is this honeytoken targeting?
2. Can the insider easily distinguish between the real and the fake document?
3. How much fake information shall the document contains?
4. Which column of information should be changed or unchanged?

Brain Bowen has proposed the use of decoy document distributor (D3) system and other sensors to detect and monitor the insider attack. The D3 system is a web-based honeypot for generating and distributing decoys document or honeytoken. The whole architecture includes decoy document distributor system which is the honeypot and, sensor such as SONAR and host level sensor.

The decoy document distributor system automatically embeds multiple signal into the decoy document to increase the probability of detection of document being misuse [28]. The signals emitted, in case of decoy document being misused, is generated by embedded honeytokens that are monitored, beacon that alert the SONAR sensor and marker which enable detection at the host level sensor. Another functionality of host level sensor is to monitor the anomalous user search action if the deviation of the file search behavior protrude largely from the baseline of normal user search behavior.

Huseyin Ulusoy proposes a method to detect attacker by the use of honey data. Honey data in reality is a form of honeytoken which is renamed by its author and is part of the honeypot trap [32]. The system is composed of three phases. The first phase is honey data generation and integration. In this phase, the honey data is generated using the actual data by the data controller. The honey data and the actual data are uploaded into the cloud. The honey data and the actual data are shuffled in such a way that each data split is formed to contain at least one honey data instant. It is then place in different position. Blacklist and whitelist are also created based upon the position information (location of honey data). The second phase is trap setting. The trap is running in kernel-mode and is used to monitor the file access system-calls. The trap send an alert message to the data controller once the position information and file request match the record in the blacklist. The last phase is map reduce job tailoring. In this phase, the system is tailored to ensure that the legitimate user does not access the honey data in the file system. Whitelist is used to ensure the above requirement is met. The honey data comes in three scale levels. They are the honey files, honey split and honey key-values.

The system ability to detect intruder by the probability of detecting an unauthorized data access which is set by the data controller and the amount of honey data content in the files is dependent on the setting of the detection rate.

Anomaly-based filter can be used to detect insider attack by analysing the pattern or flow. Such system will redirect the malicious user into a insider honeypot once the IDS detects abnormal pattern or discover deviation excesses its threshold of the normal flow. *Praveen J U* has proposed the use of misbehavior pattern detection method for IDS to detect and redirect the intrusion packet to the honeypot. The private/public keys are introduced in the proposed system to serve as a double security measure alongside with the user name and password [30].

The user has to first send an encrypted request for the ticket to a particular service from the ticket generator. Upon received the encrypted service ticket from the ticket generator, the user has to login and send the service ticket to the NIDS. The NIDS will send the ticket to the ticket manager for verification. If the verification is pass, the user will be directed to the file system else if the authentication failed, user will be blacklisted and directed to the honeypot.

Summary for insider detection honeypot: Different authors has different opinion on how to determine whether there is an insider attack. Table 3.12 reveals that framework in [30, 32] are limited to masquerader insider attack whereas for framework in [24, 28, 31] apply to both masquerader and traitor. Framework in [30, 32] assumes that the intruder is not very familiar with the system and try to access to the system. The whole concept is more efficient for detecting outsider attack rather than insider attack. Using honeypot alone to detect insider attack provide limited information as the attacker will not mingle in the honeypot for long period of time whether or not he/she achieves his/her objective. The honeytoken proposes in [28, 32] allow administrator to continue to monitor even the intruder is no longer within the parameter of the decoy system. Every action perform on the token will cause it to send feedback back to the server.

The development of the insider detection has shifted from merely using the manual honeytoken [24] to using IDS and honeypot [30] to the automated honeytoken generator system and honeypot [32]. The development of the decoy file system (honeypot) with the usage of the honeytoken file make the proposed framework in [28, 32] indistinct from the legitimate system. The concept proposed in [28, 32] make a perfect solution; however, researchers still have to face the reality of the high level of complexity in setting up the entire system.

All researchers who proposed the usage of honeytoken as part of the framework face the great challenge in making the decision for proportion of 'honeydata' and real data for the honeytoken needs to be established to prevent the trigger of the suspicious alarm by the insider in regards to the data targeted. The proportion decides is subjective and needs to rely extensively on the professional opinion on the higher management of the organization who may be highly suspicious of insider.

3.6 Advanced Persistent Threat Honeypot

Advanced persistent threat (APT) is a well-organized attack that try to gain control of the system in an organization in order to gain information for personal benefit. The attack can consists of the usage of several sophisticated attack vector or simple malware attack. The tools used in the most attack are modern state-of-the-art hacking application. It can be customized or of the shelf malware tool. The objective of APT is to remain invisible for as long as possible, move quietly from one compromised host to the next without generating regular network traffic and gain total control of the host [33].

The attack is a very low and slow process and also to ensure that the malicious activities cannot be observed by legitimate user. The intension of the attacker is similar to insider attack as not to do damage to the system but to gain unlimited access to the confidential information of the organization.

The first warnings against targeted email containing trojans to steal sensitive information are published by UK and US CERT organisation in 2005 [36]. During that time, the term "Advance Persistent Threat" is not being used. APT was later widely cited by Colonel Freg commander of the 23rd Information Operations Squadron in US Air Force.

APT attack comes in seven stages. They are:

1. The first stage is the initial compromise stage. Methods such as social engineering, spear phishing, zero-day virus and steady drive-by download from commonly visit web site are used to infect the targeted employee computer.
2. Second stage establish foothold in the targeted network. This is fulfil by using tunnelling to access the network infrastructure, install a remote administration software and create a backdoor.
3. The third stage escalate privileges to gain full control rights of the network. This is done by using exploitation and password cracking software.
4. The fourth stage is network information consolidation stage. Information such as trust relation and windows domain are gathered.
5. The fifth stage is lateral movement stage. Attacker will begin to compromise other system and server and perform data harvesting on them.
6. The sixth stage, attack requires to maintain the connection to ensure continuous control over the channel.
7. The seventh stage is also the final stage. Attacker downloads valuable data from the victim's network

The attack is often performed as a continuous process and is sometime being described using singular expression such as "the" APT attack. Current security techniques and tools find challenging to detect such attack especially stealth technique such as tunnelling is being used. Researchers in [34, 35] have proposed the use of honeypot to capture this attack.

Roman Jasek has proposed the use of multiple honeypot in several location of the network to capture APT attack [34]. The proposed system consists of high-interactive honeypot, low-interactive honeypot, honeypot for production and honeypot agent.

The honeypot agent acts like a normal human user possess the behavior of ignorance user to attract the attack. The low-interactive honeypots emit the known vulnerability into the network. The high-interactive honeypots allows attacker to interact and compromise. The act of compromising the honeypot in the honeyfarm is recorded and monitored by the administrator. Safety rules derive from the compromising act is applied to the production system.

Honeypot can be implemented next to the actual server. This creates a mirror image which can be used as a bait to attract attacker. *Zainab Saud* focuses on implementing the honeypot to protect the more important facilities such as server [35]. The proposed architecture include NIDS as a gateway to the server and low-interactive honeypot

Table 3.13 Summary of APT honeypot

S/No	Ref No	Author	Year	Framework	Interaction of honeypot	Other honeypot	Form of honeypot	Purpose of honeypot	Detection method	Nature of honeypot
3.6.1	[34]	Roman Jasek	2013	APT detection system	High-interactive	Low-Interactive	Physical	Production	Honeypot agent	Active
3.6.2	[35]	Zainab Saud	2015	APT Proactive detection honeypot	Low-interactive	Nil	Physical	Production	NIDS	Active

to emit signal to attract the attacker. The focus of this system is to ensure that the administrator receive a timely information of the intrusion once the honeypot is being compromised. The task of the NIDS is to provide a detailed picture of the attack, its log and alerts assist the administrator to analyze and correlate different events.

Summary for APT detection honeypot: In Table 3.13, the two frameworks proposed trying to solve the issue of APT in a different level. Equal scale is impossible in this case to be used for comparing the result in technological level or empirically. The framework in [34] is designed to capture a full APT attack and redirect this attack to a honeyfarm for monitoring and analytical work. The information of whole attack is being recorded and stored in a database. The framework in [35] focuses on protecting the prime facilities such as the server. It is assumed that the APT attack has already occurred and is undetected in the network.

Both frameworks adopt the passive approach to lure the attacker to the honeypot and both authors in their articles stated the belief that an advance persistent threat attack has the ability to slip through the eye of intrusion detection system and honeypot is a good solution to capture the signal of such threat [34, 35]. Both frameworks use static honeypot to emulate the vulnerability to attract APT attacker and to my dismay, there is no dynamic honeypot proposed so far to tackle APT issue.

References

1. J.P. John, F. Yu, Y. Xie, A. Krishnamurthy, M. Abadi, Heat-seeking honeypots: design and experience, in *Proceedings of the 20th International Conference on World Wide Web* (ACM, 2011), pp. 207–216
2. A. Schoeman, Amber: a zero-interaction honeypot and network enforcer with modular intelligence, in *Information Security for South Africa, 2013* (IEEE, 2013), pp. 1–7
3. S. Djanali, F. Arunanto, B.A. Pratomo, A. Baihaqi, H. Studiawan, A.M. Shiddiqi, Aggressive web application honeypot for exposing attacker's identity, in *1st International Conference on Information Technology, Computer and Electrical Engineering (ICITACEE), 2014* (IEEE, 2014), pp. 212–216
4. J. Crist, Web based attacks (SANS, 2007)
5. Imperva, Imperva's web application attack report (2011)
6. M. Akiyama, Y. Kawakoya, T. Hariu, Scalable and performance-efficient client honeypot on high interaction system, in *2012 IEEE/IPSJ 12th International Symposium on Applications and the Internet (SAINT)* (IEEE, 2012), pp. 40–50
7. R. Shukla, M. Singh, Pythonhoneymonkey: detecting malicious web urls on client side honeypot systems, in *3rd International Conference on Reliability, Infocom Technologies and Optimization (ICRITO) (Trends and Future Directions)* (IEEE, 2014), pp. 1–5
8. T.-M. Koo, H.-C. Chang, Y.-T. Hsu, H.-Y. Lin, Malicious website detection based on honeypot systems, in *2nd International Conference on Advances in Computer Science and Engineering (CSE 2013)* (Atlantis Press, 2013)
9. Y.-M. Wang, D. Beck, X. Jiang, R. Roussev, C. Verbowski, S. Chen, S. King, Automated web patrol with strider honeymonkeys, in *Proceedings of the 2006 Network and Distributed System Security Symposium* (2006), pp. 35–49
10. B.K. Mirsha, U. Kumar, G. Sahoo, Defending polymorphic worms in computer network using honeynet. Int. J. Eng. Sci. Technol. (2014)
11. S. Paul, B.K. Mishra, Honeypot-based signature generation for polymorphic worms. Int. J. Secur. Appl. **8**(6), 101–114 (2014)

12. M.M. Mohammed, E. Aleisa, N. Ventura, Zero-day polymorphic worms detection using aho-corasick algorithm
13. A.N.A. AlFraih, W. Chen, Design of a worm isolation and unknown worm monitoring system based on honeypot, in *International Conference on Logistics Engineering, Management and Computer Science (LEMCS 2014)* (Atlantis Press, 2014)
14. P. Jain, A. Sardana, Defending against internet worms using honeyfarm, in *Proceedings of the CUBE International Information Technology Conference* (ACM, 2012), pp. 795–800
15. B.K. Mirsha, U. Kumar, G. Sahoo, *Double-Sticky-Honeynet for Defending Viruses in Computer Network*, vol. 7 (2012), pp. 131–134
16. G. Portokalidis, H. Bos, Sweetbait: zero-hour worm detection and containment using honeypots. Elsevier J. Comput. Netw. (2005) (Special Issue on Security through Self-Protecting and Self-Healing Systems)
17. D. Dagon, X. Qin, G. Gu, W. Lee, J. Grizzard, J. Levine, H. Owen, Honeystat: local worm detection using honeypots, in *Recent Advances in Intrusion Detection* (Springer, 2004), pp. 39–58
18. I. Alberdi, E. Alata, V. Nicomette, P. Owezarski, M. Kaâniche, Shark: spy honeypot with advanced redirection kit, in *IEEE Workshop on Monitoring, Attack Detection and Mitigation (MonAM07)* (2007), pp. 47–52. (ps approach for preventing, detecting, and responding to ddos attacks. Br. J. Appl. Sci. Technol. **5**(5), 500, (2015))
19. S. Khattab, R. Melhem, D. Mossé, T. Znati, Honeypot back-propagation for mitigating spoofing distributed denial-of-service attacks, in *20th International on Parallel and Distributed Processing Symposium. IPDPS 2006* (IEEE, 2006), pp. 8–pp
20. R. Selvaraj, V.M. Kuthadi, T. Marwala, An effective odaids-hps approach for preventing, detecting, and responding to DDoS attacks. Br. J. Appl. Sci. Technol. **5**(5), 500 (2015)
21. S.S. Sadamate, V. Nandedkar, in *Advance Honeypot Mechanism-the Hybrid Solution for Enhancing Computer System Security with DoS*, vol. 4 (2015)
22. J. B. Grizzard, V. Sharma, C. Nunnery, B.B. Kang, D. Dagon, Peer-to-peer botnets: overview and case study, in *Proceedings of the First Conference on First Workshop on Hot Topics in Understanding Botnets* (2007), pp. 1–1
23. S.M. Khattab, C. Sangpachatanaruk, D. Mossé, R. Melhem, T. Znati, Roaming honeypots for mitigating service-level denial-of-service attacks, in *24th International Conference on Distributed Computing Systems, 2004. Proceedings* (IEEE, 2004), pp. 328–337
24. L. Spitzner, Honeytokens: the other honeypot (2003)
25. M. Bercovitch, M. Renford, L. Hasson, A. Shabtai, L. Rokach, Y. Elovici, Honeygen: an automated honeytokens generator, in *2011 IEEE International Conference on Intelligence and Security Informatics (ISI)* (IEEE, 2011), pp. 131–136
26. S. Chauhan, S. Shiwani, A honeypots based anti-phishing framework, in *2014 International Conference on Control, Instrumentation, Communication and Computational Technologies (ICCICCT)* (IEEE, 2014), pp. 618–625
27. M. Husák, J. Cegan, Phigaro: automatic phishing detection and incident response framework, in *2014 Ninth International Conference on Availability, Reliability and Security (ARES)* (IEEE, 2014) pp. 295–302
28. B.M. Bowen, M.B. Salem, A.D. Keromytis, S.J. Stolfo, Monitoring technologies for mitigating insider threats, in *Insider Threats in Cyber Security* (Springer, 2010), pp. 197–217
29. L. Spitzner, Honeypots: catching the insider threat, in *19th Annual on Computer Security Applications Conference, 2003. Proceedings* (IEEE, 2003), pp. 170–179
30. J. Praveen, P. Jayarekha, Identifying the misbehaving user in a network and trapping them using honeypot
31. L. Hong-Xia, W. Pu, Z. Jian, Y. Xiao-Qiong, Exploration on the connotation of management honeypot, in *2010 International Conference on E-Business and E-Government (ICEE)* (IEEE, 2010) pp. 1152–1155
32. H. Ulusoy, M. Kantarcioglu, B. Thuraisingham, L. Khan, Honeypot based unauthorized data access detection in mapreduce systems, in *2015 IEEE International Conference on Intelligence and Security Informatics (ISI)* (IEEE, 2015), pp. 126–131

33. P. Chen, L. Desmet, C. Huygens, A study on advanced persistent threats, in *Communications and Multimedia Security* (Springer, 2014), pp. 63–72
34. R. Jasek, M. Kolarik, T. Vymola, APT detection system using honeypots, in *Proceedings of the 13th International Conference on Applied Informatics and Communications (AIC'13)* (WSEAS Press, 2013), pp. 25–29
35. Z. Saud, M.H. Islam, Towards proactive detection of advanced persistent threat (APT) attacks using honeypots, in *Proceedings of the 8th International Conference on Security of Information and Networks* (ACM, 2015), pp. 154–157
36. E.M. Hutchins, M.J. Cloppert, R.M. Amin, Intelligence-driven computer network defense informed by analysis of adversary campaigns and intrusion kill chains. Lead. Issues Inf. Warf. Secur. Res. **1**, 80 (2011)

Chapter 4
General Purposed Honeypot Applications

Abstract This chapter will discuss about the general-purposed honeypot concept. Honeypot concept, such as the shadow honeypot, has incorporated into many honeypot framework and being described as the norm for honeypot. There are also sophisticated general purposed honeypot that can automatically adapt to the environment in to monitor different attack or able to generate a response to a human intruder. Such concept will also be discussed in-dept.

4.1 Dynamic Honeypot

One of the biggest challenges for honeypot technology is that it requires manually update and configure to adapt to the environment it is in. Configuration for honeypot is a vital ingredient for the honeypot to be blended into the environment to lure attacker towards it. Any misconfiguration will lead to the consequences such as missed detection, fail to trap attacker or compromised honeypot be used as a launch pad to launch attack on the network.

Over the years, researchers have proposed a concept called dynamic honeypot. The different between dynamic honeypot and static honeypot is its ability to automatically adapt to the environment it is in. Dynamic honeypot uses the plug-n-play concept which user can just connect it to the network and use without any configuration [6].

Dynamic honeypot does have a competitive edge when compare with its predecessor. It is able to create a honeypot which can blend very well with the environment it is in. The question is how this honeypot get the information about the environment. Researchers [1–5] use fingerprint method to collect information about its environment and the system information. Fingerprint method can be classified into two categories, namely passive method and active method.

Active fingerprint method uses active techniques such as port scanning to collect the information of the server. It sends out traffic packet to the targeted host and through the response to identify the role of each host in the network.

Xuxian Jiang proposes a catering honeypot that will actively collect information from current network traffic and dynamically create honeypots that are likely to be

© The Author(s) 2018
C. K. NG et al., *Honeypot Frameworks and their Applications:*
A New Framework, SpringerBriefs on Cyber Security Systems and Networks,
https://doi.org/10.1007/978-981-10-7739-5_4

attacked within a short span of time [1]. The architecture consists of profiler module, activation module, deactivation module, honeypot factory and deployment module.

The profile module sniffs the network to capture the activities in network. It presents the result to the activation module. The activation module will use the information to select the appropriate operating system and service, sends request to the honeypot factory to create and deploy honeypot with the selected configuration. The honeypot implemented can be high-interactive or low-interactive in nature depending on the information collected by profile module. It can also be deployed physically or virtually by the deployment module. The deactivation module isolates the honeypot in the course of total compromised and being used as a bot for DDoS attack or no more associated activities in the honeypot.

Passive fingerprint method sits quietly in the network to gather and use the information collected to identify each host in the network. It does not send traffic outwardly.

Liberios Vokorokos proposes hybrid honeypot architecture with passive fingerprint. The honeypots uses existing tools like Snort, Sebek and Dionaea. It allows deployment in any environments by auto-configure itself using the collected parameter of the selected environment through the passive fingerprinting method [2]. The selection of tool, convenient provided by the auto-configuration function and carefully selected parameter for each configuration in the tools, can help reflect the advantage of both high and low interaction honeypot and minimize the existing disadvantage.

The implementation of the hybrid honeypot is based on the client-server architecture where multiple clients are connected to a server. Data such as the event-log, malware collection and packet collection are sent to the server for storage and analysis. In essence, the author is evocating the used of the existing tools for the honeypot data collection and the main focus is on cloning the different environment into honeypot and conceal it within other production system to make it harder to detect.

Passive and active fingerprint method do have their pros and cons. The pros for active fingerprint method is

1. Less time needed to identify the host on the network
2. It is more accurate

The cons is its presence is noticeable as it is constantly admitting traffic into the network.

Passive fingerprint maintains its stealth by only sniffing the traffic in the network. However, its stealth-like cover limits

1. Its ability to produce a more accurate information about the host
2. takes longer time to produce result

In the context of dynamic honeypot, the stealth property is not prioritized, the information accuracy is given much more weight. The combine use of active and passive method help to enhance the accuracy and cover the blind spot in regard to the information of the network, host and the server.

Kartik Chawda proposes a scalable dynamic and hybrid honeypot model with passive and active fingerprint method. The model implements the low-interactive honeypot such as honeyd at the front-end capture malicious traffic and high-interactive honeypot at the backend to receive redirect traffic [4]. The hybrid honeypot is made up of six components. They are the active probing tool, passive fingerprint tool, honeypot (low-interactive and high-interactive), database, honeypot engine and administrative interface.

The honeypot engine will send the configuration to the low-interactive honeypot after gathering the required information from the probing tool and preform the necessary calculation. The honeyd intercepts the attacker once it detects a connection to an unused IP. Honeyd sends the information of the attacker activities back to the honeypot engine to be stored in the database. Scale of the hybrid honeypot is achieved when sensor filters the interaction to the high-interaction honeypot.

A dynamic honeypot should be capable of capturing multiple attack. The use of physical honeypot for dynamic concept seem to be less feasible. It is less efficient, inflexible and also incur a great cost in term of labour and hardware when a honeynet is required for the implementation. The ability to "create" multiple high-interactive honeypots to cater different attacks and "destroy" them to free up resource is vital. Virtualization tool such as VMWare, KVM and LXC have this function and can be adopted easily.

Wenjun Fan proposed a versatile virtual honeynet enabling tool that automatically configure, deploy and manage a flexible honeynet [5]. The proposed framework is comprised of six components, the sensor, the request processor, the configuration engine, the honeynet template repository, the specific translation modules and the deployment tools.

The sensor generate automated input by scanning the production network, it also compares and update the honeynet description stored. Once received the input, request processor check the request syntax. If the request syntax is correct, the configuration engine processes the request content. The configuration engine check whether the request needs. It will apply the requested template in the repository or make a complete honeynet description according to the content-input. The corresponding translation module interprets the honeynet description into specific configuration for the target deployment tool. Lastly, the deployment tools deploy the requested honeynet by processing the corresponding specific configuration.

Fingerprinting is important in all dynamic honeypot. Active fingerprinting method is used to collect OS fingerprint across the production network.

The versatile virtual honeynet framework allows multiple platform to be implemented. It uses low-interactive honeypot such as honeyd to capture automated attack and virtual high-interactive honeypot to capture human-generated attack.

The processed information is stored in a database which can be centralized or distributed. The database can be easily managed for a small network; it becomes increasingly complex and impossible to manage when the information is increasing exponentially. Traditional database can no longer manage data collected from the large scale network efficiently, especially when the data is growing in such a rapid rate. Failure to manage the data will sometime lead to failure to detect an intrusion or easily being detected. As a result, opportunity to capture value information about the attacker is lost.

Sumaiyya Khan proposes the use of Hadoop in the creation of virtual honeypot to implement the level of priority of the system to the organization. Hadoop is built to cater big data and is able distributes the datasets across clusters of servers. The proposed system used the "fire and forget" concept where the virtual honeypot is generated "on demand" or when there is an actual attack on the system using the data set collected. This helps to prevent the detection of honeypot and provides a better concealment of honeypot within the network [3]. The system has the abilities to learn and monitor the network in real time, and the honeypot emulates real service.

Summary for dynamic honeypot: Table 4.1 shows each framework uses a different approach to collect the information in the network required for the honeypot. The term used is different but the function remain unchanged. Table 4.1 also shows that all dynamic honeypot researcher use low-interactive honeypot as the prime honeypot that use for probe the attacker. This is due to the ease of configuration, however, the low-interactive honeypot do have its drawback as it is not very efficient to detect the unknown attack. References [2, 4, 5] have include both high and low-interactive honeypot in their architecture. The low-interactive honeypot, preconfigure to blend with the network, emits the vulnerability and the high-interactive honeypot will contain, monitor and store all the attack. References [2, 4, 5] do have some disadvantage as the high-interactive honeypot of the required operating system need to be preinstall so that the low-interactive honeypot can be configured to be related to it. Reference [1] have the advantage edge over all other framework as it allows the high, medium or low-interactive honeypot to be created using the honeypot factory.

The development dynamic honeypot focuses on method to gather information and effectively using those information about the network. Very few researchers has

Table 4.1 Summary of dynamic honeypot

S/No	Ref No	Author	Year	Framework	Core honeypot	Secondary honeypot	Network analysis tool and technique	Form
4.1.1	[1]	Xuxian Jiang	2004	Catering honeypot	Low, medium and High-interactive honeypot	Nil	Profile module	Virtual
4.1.2	[2]	Liberios Vokorokos	2013	Sophisticated honeypot mechanism	Low-interactive honeypot	High-interactive honeypot	Passive fingerprinting method	Virtual
4.1.3	[3]	Sumaiyya Khan	2014	Honeypot with hadoop	Low-interactive honeypot	Nil	Hadoop	Virtual
4.1.4	[4]	Kartik Chawda	2014	Dynamic and Hybrid honeypot	Low-interactive honeypot	High-interactive honeypot	Active probing tool and passive fingerprint tool	physical
4.1.5	[5]	Wenjun Fan	2016	Versatile virtual honeynet framework	Low and High-interactive honeypot	Nil	Active fingerprint tool	Virtual

extended their research to area of dynamic honeynet. The concept of virtual honeypot in [1–4] has been limited to low-interactive honeypot. Reference [5] adopted the virtual high-interactive honeynet which is capable to analyze several malware at the same time.

The early researcher uses conventional database to store network data and in the current development [3], big data application such as Hadoop is used to cater for the huge network data which conventional database no longer efficient of.

4.2 Artificial Intelligent Honeypot

Artificial Intelligent honeypot do possess the same property as the dynamic honeypot. The difference is that Artificial Intelligent honeypot has the abilities to learn and reason like a normal human being. Artificial Intelligent honeypot is able to interact with its attacker.

Artificial Intelligent honeypot also possess the ability to learn whilst interacting interact with the attacker's environment. *Adrian Pauna* proposed a honeypot system with the abilities to reinforced learning about the attacker and its environment in the event of SSH attack. The system 'learnt' by interacting with the environment uses trial-and-error mechanism [7]. This is to ensure that an appropriate policy is selected for the chosen actions in a given situations.

The system consists of a medium interactive honeypot, kippo, and the reinforcement learning module. The honeypot will interact with the attacker. The RL module will be based on the input to determine which the action module to be triggered for the honeypot to interact with the attacker. The action module comprises of five actions namely allow action, block action, fake output action, insult action and delay action. The information of the attacker can be collected successfully by allowing attacker to have limited but real interaction with the host via the proxy module.

AI honeypot also allows itself to interact with the attacker like a normal system so as to determine the attack and to allocate task to be performed next. Through interaction, knowledge is gained and procedure for the attack is formulated using certain AI algorithms. The process is learned and stored in a database. *Wari Zanoramy Ansiry Zakaria* has given a very high appraisal to self-adaptive honeypot. The proposed idea includes artificial intelligent technique into honeypot so as to possess the capabilities to resolve the real world problem.

There are two different AI techniques of the honeypot namely, case-based reasoning technique and expert system, in fact, most expert systems are rule-based. That is, the system is made up of if-then-else switches. Maybe there are thousands of rules, but system is still preliminary to build an intelligent honeypot [8]. The CBR is a way of solving problem using the past solution. CBR learns from the past case base, adapts it and reuses the solution learned to solve the similar case. The ability to learn and adapt in the CBR cycle makes it suitable for building an intelligent honeypot. The ES is a computer program that simulates the way human solves in a specific domain. It consists of working memory, knowledge-base and an inference

Table 4.2 Summary of AI honeypot

S/No	Ref No	Author	Year	Framework	Honeypot	AI technique
4.2.1	[7]	Adrian Pauna	2014	RASSH	Low-interactive	Machine learning technique reinforced learning game theory
4.2.2	[8]	Wari Zanoramy Ansiry Zakaria	2012	Intelligent honeypot	High-interactive	Case-based reasoning and expert system

engine. The working memory used the kb which contain the features of the detected production host and the IE which decide on the solution to adapt. ES can be used for the domain of honeypot configuration.

Summary for AI honeypot: In Table 4.2 shows both honeypot adapt certain type of learning theory for the honeypot to interact with the attacker. Reference [7] uses the information collected from the production network environment and use the concept of game theory to prepare the honeypot used to interact with the attacker. The framework learns and records the behavior for better interaction if same behavior encounter. Reference [8] uses the past similar case and solution and apply it to the current problem encounter to create the honeypot. The if-then rule stored in the database is used in the decision making of the content in the honeypot.

Both frameworks can become complex in the initial setup such as training the system and include the correct and relevant data. The application of artificial intelligent in the honeypot technology may be theoretically useful in the prospect of research, but it is still in its infancy stage even though AI has been around for decades. AI approaches such as soft computing and anti-logic are yet to be discovered to be introduced in any AI honeypot. Therefore, there are still a great amount of work need to be done in this area.

4.3   Shadow Honeypot

As mentioned, honeypot is not a new concept. It has been introduced at the end of last century and was then call decoy tool kit. Lance Spitzner later improved the concept and rename it as honeypot [9]. Honeypot can be used to monitor and analyze certain attack efficiently. In the beginning of this century, a new form of IDS called anomaly-based IDS has been introduced to compensate the drawback of the signature-based IDS. However both honeypot and the new IDS do have their own challenges.

Honeypot does have its limitation as some configuration is only suitable for particular type of attack. There are no one configuration for all. Anomaly-based IDS can detect both known and unknown attack, but it is sometime less accurate. The level of accuracy offers the trade-off between false positive and false negative.

A new form of honeypot called shadow honeypot is introduced. It is the combination of honeypot and anomaly-based detection.

The tasks of shadow honeypot are divided into two steps:

1. Segments suspicious traffic from the regular traffic.
2. Redirect malicious traffic to honeypot

Anagnostakis Kostas proposed a novel hybrid approaches that put together the best features of the honeypot and anomaly detection to detect attack [10]. The shadow honeypot and the actual application both share the same state or it can be defined as shadow of the real application. This is to avoid attack if the filter fails to identify the attacker. The proposed concept relies greatly upon the anomaly detection.

In this system, it uses three different types of anomaly technique such as abstract payload execution, earlybird algorithm and network-level emulation. The anomaly intrusion detection systems detect abnormal packet and the filter redirects it to the shadow instant. Once the shadow honeypot validates the attack, the filter will block to prevent it from further attack. The system is built with the feature to self-train and fine-tune itself by using verified bad traffic and known mis-prediction. The shadow honeypot can be rollback to the last known good.

The malicious behaviour traffic can also be fed to the IDS so that the IDS is able to identify the bad traffic from its abnormal pattern. *Maros Barabas* proposed a shadow honeypot system to detect and generate malicious signature for automate intrusion prevention system (AIPS) [11]. The AIDS is trained to recognize malicious activities by feeding with the packet of abnormal behavior. The dataset is capture by using three high-interactive honeypots. The high-interactive shadow honeypots with argos application simulate various system with many vulnerable services to attract attacker. They are also used to detect new thread of attack. Tcpdump collects and records the packet from the network. Communication extractor parse all data collected by Tcpdump and the honeypot. The metric extractor creates dataset of the metrics and send all relevant information with the dataset to the database. The dataset is then used by IDPS system for learning process.

Summary for shadow honeypot: One of the main ingredient for the shadow honeypot is the anomaly-based IDS. The form of IDS can be implemented using different theory. In Table 4.3, [10] has proposed the use of three different anomaly IDSs. If one IDS fail to detect the attack, the other may be able to detect it. The detection of malicious packet is identified by the knowledge of normal packet. The use of the three different forms of IDS help to reduce the level of false positive and false negative. For [11], it detects an attack by storing large varieties of abnormal behavior signature dataset.

Reference [10] does have major setback such as the time consuming process to configure three detectors to an acceptable level of detection and the log delay taken for a packet to go through the three checks. Reference [11] does also have disadvantage such as high memory usage and can cause high false positive.

Table 4.3 Summary of shadow honeypot

S/No	Ref No	Author	Year	Framework	Interaction of honeypot	Form of honeypot	Nature of honeypot	Number of anomaly IDS	Terminology use in IDS
4.3.1	[10]	Anagnostakis Kostas	2010	Shadow honeypot	High-interactive	Virtual	Production	3	Detect normal data flow
4.3.2	[11]	Maros Barabas	2012	AIPS	High-interactive	Physical	Production	1	Detect abnormal behavior data flow

References

1. X. Jiang, D. Xu, in *Baittrap: A Catering Honeypot Framework* (Purdue University, 2004)
2. L. Vokorokos, P. Fanfara, J. Radusovsky, P. Poor, Sophisticated honeypot mechanism-the autonomous hybrid solution for enhancing computer system security, in *2013 IEEE 11th International Symposium on Applied Machine Intelligence and Informatics (SAMI)* (IEEE, 2013), pp. 41–46
3. D.D.M. Khan, Z Sumaiyya, R.L. Pardhi, in *A Review on Creation of Dynamic Virtual Honeypots using Hadoop*, vol. 5 (2014), p. 421
4. K. Chawda, A.D. Patel, Dynamic & hybrid honeypot model for scalable network monitoring, in *2014 International Conference on Information Communication and Embedded Systems (ICICES)* (IEEE, 2014), pp. 1–5
5. W. Fan, D. Fernández, Z. Du, Versatile virtual honeynet management framework. IET Inf. Secur. (2016)
6. Lance Spitzner, 2003, Dynamic Honeypots, www.securityfocus.com/infocus/1731
7. A. Pauna, I. Bica, RASSH-reinforced adaptive SSH honeypot, in *2014 10th International Conference on Communications (COMM)* (IEEE, 2014), pp. 1–6
8. W.Z.A. Zakaria, M.L.M. Kiah, A review on artificial intelligence techniques for developing intelligent honeypot, in *2012 8th International Conference on Computing Technology and Information Management (ICCM)*, vol. 2 (IEEE, 2012), pp. 696–701
9. L. Spitzner, The value of honeypots, part one: definitions and values of honeypots (2001)
10. K.G. Anagnostakis, S. Sidiroglou, P. Akritidis, M. Polychronakis, A.D. Keromytis, E.P. Markatos, Shadow honeypots. Int. J. Comput. Netw. Secur. **2**(9), 1–15 (2010)
11. M. Barabas, M. Drozd, P. Hanáček, Behavioral signature generation using shadow honeypot. World Acad. Sci. Eng. Technol. **2012**(65), 829–833 (2012)

Chapter 5
Other Honeypot Applications

Abstract This chapter addresses current available concealment techniques for honeypot, and applications of honeypot for security and forensics.

5.1 Concealment of Honeypot

One of the characteristics of honeypot is its high level of concealment which helps to prevent attacker from distinguishing the legitimate system. The concealment of the honeypot can be determined by its vulnerability level and the knowledge of its attacker. For example, a system with multiple vulnerabilities or several ports open can be interpreted as a honeypot by an expert attacker. It can also be treated as legitimate system by novice hacker.

Most of the researchers are attack focus and have put all their ounce in the following:

1. Appropriate setup to capture viruses, worms and attacks
2. Capture attacker's activities by releasing popular vulnerabilities

One of the greatness of honeypot is to lure attacker into it without realizing he/she is in a honeypot. The concealment of honeypot plays a significant role in the decision for the attacker who decide to attack the system. Its purpose is defeated if attacker detects he /she is in a honeypot and detracts from it. Researchers in this area have derived the following solution:

1. Fake honeypot and fake host tactic: One of the simple but effective methods to conceal honeypot within the real hosts is to make honeypot identical to real hosts or configure the real hosts to emit honeypot signal [1].
2. Real host bait tactic: Use real server as a bait to lure attacker to compromise the system [3]. Honeypot is activated to receive incoming packet once malicious traffic signal is detected.
3. Conceal honeypot application in system kernel tactic. This method is long been used by high interactive honeypot such as Sebek this application is dependence to the some of the feature libraries in the operating system [4].

© The Author(s) 2018
C. K. NG et al., *Honeypot Frameworks and their Applications:
A New Framework*, SpringerBriefs on Cyber Security Systems and Networks,
https://doi.org/10.1007/978-981-10-7739-5_5

The fake honeypot and fake host tactic allows the honeypot will receive timely update to its software and operation system. This will create a psychological tension to the attacker so that he/she will relentlessly attack the host base on the wrong judgement. The former is the real host actively emitting its identity as an obvious honeypot to avoid being attacked. *Leyi Shi* proposed the concept of protective coloration and warning coloration for the honeypot. The concept was derived from the idea of fraudulent countermeasures use in nature to avoid being discovered. The model of the honeypot adapts a five-tuple namely, the mimic, the model, the dupe, the status of service and honeypot, and the evolution [1]. The protective coloration is used to attract attacker to attack the service and the warming coloration tries to avoid attack.

The honeypot is implemented and conceal among the real host with the real and updated configuration. This is to lure the attacker to attack the honeypot assuming that it is the real host. This is the process of protective coloration. The warming coloration is fulfilled by server resemble the characteristic of a honeypot to evade the attention of the attacker. The system is able to adapt to the change of current network environment via launching the evolution process once its identity is being revealed.

The real host bait tactic hides honeypot behind the host or server it is representing. The concept is proposed by *Daisuke Miyamoto* to improve the presence so that the honeypot is not easily noticeable by the attacker. He proposes a migration technique called INTERCEPT to duplicate the actual web server and isolate the attack to a virtual high-interaction server type web-based honeypot [3]. The actual system and the honeypot share the same IP and MAC address. In order to differentiate malicious user from legitimate user, heuristics-based web application firewall is used to calculate the likelihood of attack in the suspicious request, then software defined network such as openflow or LISP are used to redirect the suspicious traffic to the honeypot.

In the process of migration, the author mentions two challenges which need to be considered. There are the timing and the data issue. The process of cloning the web server, distinguish of the malicious connection and the redirection process need to occur within an acceptable timing to avoid the suspicious of the attacker. The system also needs to ensure that the data is not leaked or corrupted while in the process of migration.

Christoph Pohl briefly describes the important factors that contribute to a good high-interactive honeypot and proposed a kernel module honeypot tool called Apate for the high-interaction Linux honeypot [4]. A high-interaction honeypot should be able to be hidden from its attacker, exploitable and at the same time prevent being taken over by attacker to cause harmful act to others.

Apate possesses the above mentioned properties and has even exceed in term of the performance and functionalities of its predecessor, Sebek. Furthermore, Apate is able to perform fine-grain information logging, manipulate the behavior of system calls and low computational overhead [4]. Apate follows these three major objectives which are the reasons that made this kernel module honeypot tool more preferable over others. First, it provides highly flexible configuration which ensure every possible ruleset and action is covered. Second, apate achieves a high level of invisibility that it has hidden itself from common utilities tool and also provides no trace in common directory and log entry. Third, Apate is able to serve under productive usage scenarios.

Summary for concealment of honeypot: Table 5.1 reveals the method of how honeypot conceal themselves from being discovered. The framework in [1] use the method similar to the fake honeypot concept [2]. The honeypot will emit the properties of the real host into the network and the real host will do the same. The framework in [3] uses the real host as a bait to lure the attacker to attack. It relies on the filter to determine the malicious attack. Within a short period of time, a honeypot is generated to contain the attacker. The honeypot in this framework is concealed behind the actual system and in the IP prospective, the attacker is unable to distinguish as it shares the IP address with the actual system. The framework in [4] conceals its honeypot operation in kernel level where it is not reveal even if the attacker possess administrator rights. The system does not store log file in the common directory.

The framework in [1] has an excellence concealment for honeypot, but the concealment concept provides relatively poor security measures which defeat the purpose of honeypot for production. The framework in [3] uses the real system to tempt the attacker and redirect the traffic to the honeypot. The interpretation of malicious traffic relies solely on a single firewall implemented which imply the lack of emphasis on the traffic check mechanism. The overall concealment concept is excellence and provide great flexibility for further modification. The framework in [4], like the Sebek kernel based honeypot, can be easily disable if the attacker install malware which reset the kernel configuration.

5.2 Application of Forensic in Honeypot

Forensic comes from the latin term, forensis, means "of or before the forum". The process of forensics is used to preserve and present evident in the court of law to press charges against individual. It can also be used in information technology realm to get a full view about the intruder.

Digital forensics can be divided into two sub-disciplines. They are computer forensic and network forensics [5]. In computer forensics, the analytical investigation is primarily performed on the digital storage media such as volatile memory and hard disk. The network forensics focuses its investigation on analyzing the traffic packet for abnormal pattern.

Honeypot provides a wealth of information about the intrusion and it provides useful and relevant data for the work of forensics. It is important to make the distinction between traditional digital forensic approach and the forensic approach for honeypot [6]. It is paper, we will use honeypot forensics to replace forensic approach for honeypot or forensic application for honeypot. Honeypot forensic is specially tailored for the framework of honeypot and it possesses the following characteristics:

Table 5.1 Summary of honeypot concealment technique

S/No	Ref No	Author	Year	Framework	Interaction of honeypot	Form of honeypot	Purpose	Conceal method	IP address
5.1.1	[1]	Leyi Shi	2012	Mimicking Honeypot	High-interactive	Physical	Production	Protective coloration	Separate IP address
5.1.2	[3]	Daisuke Miyamoto	2014	Intercept	High-interactive	Virtual	Production	Duplicate actual system	Same IP address as the server
5.1.3	[4]	Christoph Pohl	2015	Apate	High-interactive	Physical	Production	Kernel level Configuration	Separate IP address

1. Analyse lesser but informative traffic packet
2. All incoming traffic are malicious
3. All activities are considered bogus.

The above characteristics make honeypot forensic less complex.

Like the traditional digital forensics work, careful planning and selection of appropriate tool for honeypot forensic are essential to successfully identify the event and motivation of the intrusion. The type and setup of honeypot and its environment need to be part of the consideration as it will affect the investigation [19]. Benjamin Franklin quotes,"If you fail to plan, you are planning to fail!" [8].

Researchers have proposed different honeypot forensic methodologies or frameworks and architectures to assist honeypot forensic to investigate general attack such as malware or specialized attack such as botnet. Table 5.2 provides a belief summary of the method and framework proposed for honeypot forensic.

In this section, the honeypot forensic will be focused on the two categories of attacks namely general type attack and botnet.

5.2.1 Honeypot Forensic for General Type Attack

The honeypot forensic approach for general attack does have its challenges as it needs to derive a "one method fit all" framework so that no attack will be missed by the forensic investigator in the forensic investigating process. But this is not always the case.

Honeypot forensics is divided into two sub-categories and they are the main processes use to investigate the honeypot.

Both processes are important as they complement the drawback of each other. Example, heartbleed attack leaves no trace in its host while retrieving sensitive information such as personal banking detail from the volatile memory. Computer forensic is unable to discover such attack [9]. Network forensic is able to trace such attack by analysing the abnormal traffic from the compromised host.

However, there are some researchers [10, 11] believe that network forensic solely is adequate to solve the mystery of the intrusion.

One of the method in the network forensic is to group related traffic together. This helps the investigator to easily identify the malicious apart from the legitimate traffic. This is not an issue in honeypot network forensic as all traffic activities in honeypot network is deemed malicious. The grouping helps to differentiate attack pattern. *Pouget F.* has proposed a clustering approach when applying forensic on the honeypot data. The data of repetitive attack may create a misleading outcome without an in-depth analysis using various processes [10]. Clustering method divides the packet into groups according to its regularity. Each cluster can represent a single process or a group of processes.

In order to ensure that such clustering approach does produce performance which fulfil high level of satisfaction, it is necessary to identify and formulas a set of

Table 5.2 Summary of proposed forensic method

S/No	Author	Year	Framework	Description	Purpose
5.2.0.1.1	Pouget F.	2004	Clustering approach	Group related process according to its regularity into cluster	General
5.2.0.1.2	Alec Yasinsac	2002	Parallel architecture	Technique proposes to collect and analyse evident in order to bind intruder legally	General
5.2.0.1.3	Frederic Raynal	2004	Three phases honeypot forensic	Three phases including network analysis, system and files analysis and evidence gathering	General
5.2.0.1.4	Kara Nance	2009	Virtual machine introspection	Forensic framework for VM honeypot	General
5.2.0.1.5	Deepa Srinvasan	2012	Time-traveling forensic analysis framework	Construct a timeline for forensic investigation	General
5.2.0.1.6	Syed Ahmad	2013	Virtual HIH monitor	VMI adopts TOMI that help to identify hidden malware in honeypot forensic process	General
5.2.0.1.7	Mario Golling	2015	Decentralized approach for geolocation pre-incedent forensic	Forensic layer and honeypot or production layer is separated. Geolocation is used to gather user information	General
5.2.0.1.8	Xianming Zhong	2015	Novel virtualization based monitoring system	VMM in driver form to minimise the chance of being compromised and maximise the accuracy of presented information for live forensic	General
5.2.0.1.9	Zhenxin Zhan	2013	First statistical forensic framework base on the novel concept of stochastic cyber attack process	"Gray-box" prediction process to predict incoming attack	General
5.2.1.1	Sven Krasser	2005	Graphical Aided forensic method	Use visualization aid to identify malicious activity	Botnet
5.2.1.2	Van-Hua Pham	2011	Grouping technique	Group attacks according to their location and OS platform	Botnet
5.2.1.3	Brian Cusack	2014	Five levels forensic method	Traffic is focused on different analysis in each level.	Botnet

terminology to find the root cause by performing an in-depth analysis against the honeypot data [10]. The first goal is to identify the attack source IP, number of machine attacked, number of packet sent to each machine and its ports sequences. Then the application of association rules which help to address the unique data mining issues. This technique helps investigator to uncover hidden pattern in the dataset by looking for the similar regularity. Last is to determine the levenshteins distances for each cluster to ensure each cluster is not miscategorised.

Network analysis unlike other forensic work, it deals with volatile and dynamic information. The traffic packet in the network is lost after transmission. Hence, network forensics is often referred to as a pro-active process.

Mario Golling proposes decentralized approach for geolocation-based network forensics to investigate the data capture in the high-interactive honeypot. It focuses on how the framework conducts the pre-incident preparation [11]. The concept segregates the production system and the relevant forensic component. The framework consists of four components namely caching service, geolocation service, manager and raspberry PIs.

Raspberry PI is equipped with application such as syslog and flow collector to capture and store packet traffic and system configuration information mined from the production and forensic network. Caching service is responsible for providing transparency in the control of communication between the components in the forensic layer, system updates and ability to access to information outside of the network. Geo-location service extracts and provides detailed information of the malicious IP address stored in the geo-location database. Manager is responsible to evaluate on the incident reported by the raspberry PI and information received from external source, determines the level of suspiciousness and stores the result in geo-location database as geo-reputation.

The level of suspiciousness, its name suggests, is a matrix used to determine the level of anticipated damage to the system and is formulated based on the following factors, the quality of the information obtained, blacklisted IP address, suspiciousness of the country, suspiciousness of ISP provider, a temporal element, CVSS and dynamic geo-correlation.

While network forensics identifies and analyses information from the network packet, computer forensic identifies, recovers, analyzing and presenting facts of the attack based on the information resided in the honeypot. Computer forensics allows its investigator to view and understand the full process and the destructive capability of a exploitation in the honeypot.

In the event of an attack, digital investigator looks for sign of intrusion by analysing the log, application such as email and physical and volatile memory. Researcher such as *Kevin D. Fairbanks* looks to find trace of intrusion by analysing some of the properties in the file. He proposes a timekeeper system that extract the metadata information from the honeypot and storing it for forensic analysis [12]. The proposed system is able to tracks modification, access, and change times in a way that make malicious modification of the file system difficult to hide. The timekeeper system has two basic functionalities namely extraction of Ext3 and journaling. Journaling process acts as a circular log that does or does not replay file system operations to bring the system to a consistent state within a short period of time.

The timekeeper does have other function such as build a profile of what files are regularly touched, determine irregularities and compare the inode times lifted from the system. The proposed architecture provides flexibility to incorporate with other forensic tool and it is not meant to be a stand alone system which is the setback. Third, external application such as deep security from Trend Micro can be used to monitor the attack inside honeypot with the attacker notice it.

So far, we have discussed the honeypot forensic method proposed by researcher for high-interactive honeypot and there are very little focus on the low-interactive honeypot. This is due to the reason that the pay-loads are often not available for analysis work. This, however, does not stop researcher [13] to investigate and retrieve vital information about the attack. *Zhenxin Zhan* has proposed the statistical forensic framework which is built on the novel concept of stochastic cyber attack process to analyse low-interactive honeypot data [13]. This framework can also be used for high-interactive honeypot which contains richer information. The concept can be used to represent at network-level, victim-level and port-level. The framework is capable to predict incoming attack base on the exhibited long-range dependence (LRD) in the "gray-box" prediction.

This framework comprises five steps namely data preprocessing, basic statistical analysis, advanced statistical analysis, exploiting the statistical properties and exploring cause of the statistical properties. The data preprocessing step divides the attack data flow into production and non-production. The basic statistical analysis often try to capture two properties from the attack data which can be used for advanced statistical analysis. There are attack rate (number of attacks per unit time) and attack inter-arrival time (time between two consecutive attack). Advanced statistical analysis analyse the number of attacks occur during the time intervals. Exploiting the statistical properties uses the "gray-box" prediction on the attack rate. In exploring the cause of the statistical properties, researcher tries to find out source for causing the statistical properties.

As the number of organisation moving their server into the cloud service is increasing. Computer forensic for virtual honeypot has received an enormous amount of spotlight. There are several advantages when comparing computer forensic performs virtually or physically. First, virtual investigation allow researcher to view and retrieve information needed in the middle of the attack. Second, researcher can extract the memory of the honeypot without creating changes to the environment. Third, the compromised honeypot can be access externally using agent.

Kara Nance has investigated on the implication of virtual machine introspection (VMI) which is first proposed by Garfinkel [14] for digital forensic analysis. This analysis focused is only applicable to virtual honeypot [15]. Virtual machine introspection allows system administrator to "look into" the attacked virtual honeypot without executing any application in the attacked VM. The attacked VM can be accessed with virtual machine monitor or second VM running under the control of VMM.

The VMI resolves the issues for non-quiescent analysis such as observer effect, authenticity of the result produced from the compromised system utilities in question and ease of detecting the trace of analysis by attacker. The issues mentions above seriously obstruct and affect the accuracy of the analytical work performed in progress. The VMI also overcomes the limitations of quiescent analysis loss of memory content, deceive data obtained from compromised system library and consistency memory problem causes by location that not yet being read is changed.

Virtual machine introspection itself do posed several issues. This includes:

1. VMI development tools itself
2. Application to Non-quiescent VMs
3. VMI convert operations
4. VMI detection

The security of the VMI tool such as VMM is vital as it will cause catastrophic consequence if it is compromised. The question such as how to conceal it from being detected has been the focus point for all VMI researcher. The application of VMI to non-quiescent VMs does posed the question of how this tool can be implemented into non-quiescent VMs without compromising the performance of VMI and also at the same time overcome the limitation mentioned. VMI has a great potential to create no data or modify existing data. The question such as whether the VMI has the covert operation that protect itself from insider attack or malicious VMM administrator or malicious remote outsider attack. Can the VM detected be easily convinced as a legitimate host by the attacker? These questions provides ample rooms to assist researcher to develop a better forensic tool to accurately analyse the activities of the attacker.

Xianming Zhong is able to overcome the setbacks in [15] by proposing a novel virtualization based monitoring system for live computer forensic [16]. The proposed framework, VAIL possesses the following qualities. They are:

1. VAIL is loaded as an OS driver to build the VMM layer, it requires no binary modification to the existing application and the guest operating system
2. its code size is relatively small which prevent it from being target of any attack.

With the help of hardware assisted virtualization, the framework adopts silent-virtualization technique that enable itself to build the hypervisor layer silently in the background. This also allows VAIL to be loaded even in the mid of a running guest system. VAIL is able to clear its memory footprints by adopting the memory-hiding mechanism. The memory-hiding mechanism comprised of two aspect of integrities namely static integrity and runtime integrity helps to prevent the code of VAIL to be modified and hide itself by creating and restrict the access of the private page which will not be seen by the intruder.

The improved architecture on the existing framework enables high accurate information about the guest system such as CPU state, the physical memory content and the I/O activities to be obtained for forensic even if the system kernel is being compromised.

Virtual honeypot can be implemented in two different software application. One is VMWare and the other is in Xen. Despite VMWare is the leading expert in virtualization application, the free version only provides limited feature. Xen is an open-source and all advanced options are available to be used for free. This reason, xen earns its increasing popularity. *Syed Faraz Ahmad* has proposed a virtual high-interactive honeypot monitor which uses virtual machine introspection on Xen. The architecture extracts the information from the compromised guest's volatile memory and parse the result into a log store in the server for later analytical process [17]. The system comprises of low and high interactive honeypot, honeybrid server, forensic analysis engine (FAE) and virtual machine introspection (honeymon).

The high interactive honeypot allows itself to be interacted and compromised by the intruder. The honeybrid server stores data collected in the honeypots. The forensic analysis engine extracts, processes and stores the forensic information from the log in the server, and change the privilege of the document to prevent accidental or deliberate amendment or deletion by the administrator or intruder. The entire architecture adopts the triple point memory introspection (TPMI) technique which extracts the memory, generate signature of the memory and evaluate the free space. The TPMI allows the malware to be surfaced even if it is hidden within free memory space or by modifying the memory structure.

Virtual machine for honeypot does have one unique feature, snapshot. Snapshot before and after the attack can be taken so that researcher can replay the attack. *Deepa Srinvasan* has proposed a time-traveling forensic analysis framework to investigate the attack for high-interactive virtual honeypot. The proposed analysis allows timescope that applies and extends the captured record and replay to high-interactive honeypot [18]. The framework is operated in two different modes; virtual machine record logs the activities of the honeypot and takes snapshot periodically and virtual machine replay from the selected snapshot point.

The architecture of the framework consists of four modules namely contamination graph generator, transient evidence recoverer, shellcode extractor and break-in reconstructor. In order to accurately capture the full replay of the intrusion without causing perturb to the deterministic session of the replay, all these modules are implemented outside the virtual honeypot.

Contamination graph generator provides a high-level view of attackers' behaviour in graphical image. This module typically activated immediately after a suspicious detection point in the honeypot has been identified. Transient evidence recoverer recovers attack evidence that may be erased. Given the starting and ending point of the snapshot, this module copies all the captured write file activities to the host file system. Shellcode extractor module identifies and extracts the shellcode with is not save onto the disk from the memory. Break-in re-constructor is to perform fine-grained analysis to understand how the execution of malicious, injected code hijacks control flow and tampers some of the system resources and kernel memory. This module generates an instruction execution trace.

Traditional method such as network and computer forensic can be used concurrently to assist the work in honeypot forensic. *Frederic Raynal* has proposed that the honeypot forensic can be divided into three phases, namely network activity analysis, system & files analysis and evidences gathering phase [19]. Before elaborate in detail each phase, it is essential to know the context of honeypot. The steps comprise of knowing your honeynet, knowing your network, knowing your system and lastly knowing your enemy.

Network analysis gathers the network activities from firewall logs, network flow from network device and network packet capture from inline and passive network sensor; and then extract and compile the selected information into meaningful visual representation. Next is to create and select the relevant network flow to the analysis so as to pinpoint the vital event which will be used to build the timeline.

System and files analysis provides detailed information about the activities within the system. Activities includes rootkit used, script used, binaries installed will be analysed base on the log information collected from the system logs, monitoring tools and honeypot application such as Sebek. These information will then incorporate to form a complete timeline.

Evident gathering phase provides a full picture of the motive of the attacker, identity of the attacker and malware used base upon the timeline compiled in the previous phase.

The honeypot forensics does have a major setback. The system administrator or honeypot owner using honeypot as a bait to lure intruder is not able to use the information collected from honeypot as evident in the court of law to file a law suite against the intruder. The reason behind is that honeypot purposefully emit system vulnerability which arose and tempt a "reasonable" person to commit an intrusion act. In simple term, intruder is not liable for any damage he/ she done to the honeypot. This also means that the law will be turning against the honeypot owner and person who use them [20].

Honeypot owner and system administrator need to be aware of:

1. The honeypot owner will be liable for the attacker's action in any consequences of the compromised honeytrap.
2. The intrusion will not consider a criminal act as it is deliberately allow to do so.
3. The honeypot does not contain valuable information which has not economical impact or real damage, hence their activity will not unlikely consider a crime.
4. The honeypot has to deal with the issue of privacy.

Researcher likes *Alec Yasinsac* has proposed the concept of parallel architecture and forensic method to challenge such saying. The proposed system is setup in such a way that it overcomes the above mentioned ground rules and ensure that legal proceeding is possible if intrusion occurs.

The architecture comprises of two systems namely the honeytrap system and production system [20]. The aims is to provide a framework that able to bind the intruder legally and to be free from any legal proceeding which may be to the disadvantage of honeypot owner. Both honeytrap and production system work independently from each other. The honeytrap attracts the attacker to attack honeypot, hoping that the same attacker will also attack the production system.

The proposed architecture does need a new forensic approach for the honeytrap and for the production network so as to fulfill the task being assigned to the parallel architecture. The traditional computer and network forensic comprises of scenario such as attack occur, detection of attack, last information gathered and verdict/settlement. The new forensic method for is similar to the traditional CNF method, the major difference is that the third stage of new forensic investigation is analytical work which allow the forensic process to be continuous even in the mid of intrusion and present evidence [20].

Summary for honeypot forensic for general attack: Researchers have proposed several forensic method to investigate general attack using network forensic, computer forensic and both computer and network forensic. Theoretically, the proposed framework seen to have achieved excellence outcome. Some of the frameworks does have drawback which we will outline below:

1. The clustering method [10] provides an efficient way to classify each attack based on the properties mentioned above, the knowledge of the attack relies on the availability of information stored in the database. This can be easily overcome by collecting information via tools fingerprinting from network environment or through advisories pages. The other disadvantages of this approach includes its time and labour intensive process as each process requires manual observation and handling.
2. For geolocation-based network forensics [11], its database relies heavily on available information collected online and forensics method will fail in the absent of information of the unknown such as new unknown attack source IP, suspiciousness of the country, ISP provider and dynamic geo-correlation.
3. Timekeeper proposed does have setback [12]. This forensic method is suitable for newer operating system which has Ext3 preinstalled. The "older" version honeypot may not be compatible.
4. The parallel architecture [20] does have a few major flaws. If the attacker does not have strong desire to comprise all host within the network, the system will defeat its purpose. The system is also difficult to relate the attacker in the honeytrap and the attacker in the production system. It is impossible to be sure that the person attack the honeytrap is the same person compromising the host in production network. The system does not provide method to identify the attacker in the production network.

5.2.2   Honeypot Forensics for Botnet

In this section, we will focus on Botnet attack which is widely used by hacker to mass distribution of malwares, illegal activities or attacks. Botnet can be classified into two distinct model, client-server model and peer-to-peer.

The first model consists of a centralized command centre and a group of zombie client. The command centre send instruction directly via IRC channel or http to the client. The clients in this model have no communication with each other. The client-server botnet can be easily detected and taken down by authority, hacker has moved towards P2P as an alternative.

The peer-to-peer botnet does not have a central command centre. The peer share resources among each other. The advantage is that the command centre can be switched among the clients within the botnet. The detection of such botnet is much more tedious when compare to client server model.

The purpose of exploiting a host is to collect a group of computer to form a network. This is called recruitment. The exploitation such as creating a backdoor for IRC channel to be opened up so that communication can be established. These malwares can be detected by performing a network analysis and its activities can be discovered by closely examine the memory file in the process of computer analysis.

Brian Cusack has proposed a five levels forensic analysis technique to investigate the presence of botnet capture in honeypot. The five levels forensic analysis technique covers signature analysis, binary execution and observation, network analysis, interception of C&C communication and breach of botnet [21].

In the first level, signatures and attack vector of the malware are extracted by the outsourcing agent. The second level involves binary execution and observation in a static test environment to gain knowledge of the malware behavior through the reconstruction of the event. The network observation in level three is used to capture the botnet behaviour in the network. The fourth level includes locating the IP source of the controller and the IP address of other zombie host which will eventually crease the C&C communication and disrupt the botnet. The task of breach and destroy the botnet in the fifth level helps to return the control back to the legitimate internet user.

The five levels forensic technique includes traditional forensic methods such as network forensic and computer forensic to ensure it does not exclude any area which will subsequently become a blind spot for the investigator. The technique also relies on a number of open source security and forensic software such as Helix Pro, Volatility framework, virus total and CW sandbox, and outsourcing to produce report and detection of bot intrusion.

Botnet does emit a specific traffic pattern in the network. Once the host has been compromised and control by the bot master, the bot master will test it to ensure it is truly in his/her control and use. For the client server architecture botnet, the host will be response once the C&C server sends a request. In P2P architecture botnet, the compromised host will actively engage with its peers to show its existent and also to ensure that they are still "alive".

Sven Krasser has adopted network forensic to observe this pattern for botnet. Reference [21] has proposed the use of visualization to present the traffic capture in the honeypot. The graphical presentation is represented by the use of tightly-coupled, animated, time-sequence scatter plot and parallel coordinate plots in two dimensional and three dimensional [22]. The proposed method is equally efficient in displaying the information for post-incident and "in the middle" incident by inputting the live packet data and previously captured data. The system also provides great flexibility for investigator to magnify certain point to pan the view in the graphical result.

The advantage of this proposed framework is that it allows investigator to identify malicious activity via identification of the distinctive patterns at ease especially botnet traffic. It also greatly reduces the time required for forensic work.

The "real" host (honeypot) may be eyed by more than one bot masters from the wild. Different botnet master has different motive and uses different application which will create a different but consistent pattern from exploit to control the host. *Van-Hau Pham* proposes a network and computer forensic method to identify and group together trace collected in the low-interactive honeypot. He has observed and discovered that the same botnet may be used by its bot master to attack the same host [23]. The method consists of three portions namely, attack event detection, identify action set and the characteristics of the zombie armies.

Attack event represents the existence of coordinated attack by a group of zombie hosts. The group is divided into two different categories. It is identified by the country the zombie host is resided and the other is by the operating platform used. The first portion is to identify the country, platform and cluster of the zombie hosts recruited by the bot master and assign them to a group accordingly.

An action set is a group of attack events which are highly probable the result of same group of botnet. In the second portion, the graphical representation for zombie host grouped according to its category are compared to identify the appropriate grouping which will more accurately pin point the correct group. The singleton set will be excluded from the analysis.

The information computes in first and second portions also assist in identify the main characteristics for the group of zombie hosts such as the lifetime of zombie army, lifetime of the zombie host in the botnet and the attack capacity.

Summary for honeypot forensic for Botnet: References [21, 23] have proposed honeypot forensic framework to investigate botnet by using CNF or graphical representation.

Reference [21] seems to have several blind spots in the graphical representation. The graph is an output compilation of several features from the network. The representation is only useful for known botnet pattern. The compilation for some botnet might be the same as the legitimate network pattern. It does not prompt the user of new unknown botnet if the malicious pattern match the baseline of legitimate pattern.

5.3   Direct Role of Honeypot in a Security System

So far, we have discussed the explicit use of honeypot to the security application. The contribution of honeypot enable us to understand the method and tools used, motivation and level of skill of the intruder. In this section, the implication of honeypot directly in a security system will be discussed.

According to Lance Spitzner [25], "these honeypots do not add direct value to a specific organization; instead, they are used to gather intelligence on the general threats organizations may face, allowing the organization to better protect against those threats", this is true when honeypot is used as a system that allow attacker to compromised. Honeypot can be used as part of the security system not to contain the attack but as a preventive measure to dealt with the attack (Table 5.3).

Table 5.3 Summary of honeypot in a security system

S/No	Framework	Ref No	Author	Year	Interaction of honeypot	Form of honeypot	Task of honeypot	Detection method
5.3.1	Web application honeypot	[24]	Ashwini Pawar	2014	Low-interactive	Physical	Filter and record attack	Low-interactive honeypot

This probability of such usage is low and is an uncommon sight, however, it shows that it is possible for honeypot to be directly involved in a security system. *Ashwini Pawar* uses honeypot as the main core in the defend system to prevent the production network from being intruded [24]. The system architecture consists of a two port router connected to a twenty-four ports switch, a low-interactive honeypot connected to the switch, http server and telnet server connected to the router and honeypot.

The traffic enters via the router to the switch which is configured to direct the packet to the honeypot core rather than the production network. The honeypot process and log the packet before directing them to the http server. The http server checks for malicious IP and perform the normal http function. The http server uses a filter to detect malicious IP by determine its source, destination, unused IP address and illegal combination of TCP flag or illegal content in the UDP header.

Summary for Direct Role of honeypot in a Security System: Such application of honeypot shows, but not necessary, the limitless use of honeypot. This, however, does not enhance the initial objective for the honeypot. The role of the honeypot [24] remains unchanged. The question needs to be asked is that can we use a log mechanism and NIDS to replace honeypot for more effective logging function, if it is merely use for that reason.

References

1. L. Shi, L. Jiang, D. Liu, X. Han, Mimicry honeypot: a brief introduction, in *2012 8th International Conference on Wireless Communications, Networking and Mobile Computing (WiCOM)* (IEEE, 2012) pp. 1–4
2. N.C. Rowe, B.T. Duong, and E.J. Custy, Fake honeypots: a defensive tactic for cyberspace, Information Assurance Workshop, 2006 IEEE, 2006, pp 223–230
3. D. Miyamoto, S. Teramura, M. Nakayama, Intercept: high-interaction server-type honeypot based on live migration, in *Proceedings of the 7th International ICST Conference on Simulation Tools and Techniques* (ICST (Institute for Computer Sciences, Social-Informatics and Telecommunications Engineering), 2014), pp. 147–152
4. C. Pohl, M. Meier, H.-J. Hof, Apate-a linux kernel module for high interaction honeypots (2015), arXiv preprint arXiv:1507.03117
5. E.S. Pilli, R.C. Joshi, R. Niyogi, Network forensic frameworks: survey and research challenges. Digit. Investig. **7**(1), 14–27 (2010)
6. Q. Nasir, Z.A. Al-Mousa, Honeypots aiding network forensics: challenges and notions. J. Commun. **8**

7. F. Raynal, Y. Berthier, P. Biondi, D. Kaminsky, Honeypot forensics, in *Proceedings from the Fifth Annual IEEE SMC Information Assurance Workshop, 2004* (IEEE, 2004), pp. 22–29
8. B. Franklin, Azquotes.com
9. Z. Durumeric, J. Kasten, D. Adrian, J.A. Halderman, M. Bailey, F. Li, N. Weaver, J. Amann, J. Beekman, M. Payer et al. The matter of heartbleed, in *Proceedings of the 2014 Conference on Internet Measurement Conference* (ACM, 2014), pp. 475–488
10. F. Pouget, M. Dacie et al. Honeypot-based forensics, in *AusCERT Asia Pacific Information Technology Security Conference* (2004)
11. R. Koch, M. Golling, L. Stiemert, V. Eiseler, F. Tietze, G.D. Rodosek, A decentralized framework for geolocation-based pre-incident network forensics, in *2015 IEEE 7th International Symposium on Cyberspace Safety and Security (CSS) High Performance Computing and Communications (HPCC)* (IEEE, 2015), pp. 1210–1218. (2015 IEEE 12th International Conference on Embedded Software and Systems (ICESS), 2015 IEEE 17th International Conference on)
12. K.D. Fairbanks, C.P. Lee, Y.H. Xia, H.L. Owen III, Timekeeper: a metadata archiving method for honeypot forensics, in *IEEE SMC Information Assurance and Security Workshop, 2007 IAW'07* (IEEE, 2007), pp. 114–118
13. Z. Zhan, M. Xu, S. Xu, Characterizing honeypot-captured cyber attacks: Statistical framework and case study. IEEE Trans. Inform. Foren. Sec. **8**(11), 1775–1789 (2013)
14. T. Garfinkel, M. Rosenblum et al., A virtual machine introspection based architecture for intrusion detection. NDSS **3**, 191–206 (2003)
15. K. Nance, M. Bishop, B. Hay, Investigating the implications of virtual machine introspection for digital forensics, in *International Conference on Availability, Reliability and Security, 2009 ARES'09* (IEEE, 2009), pp. 1024–1029
16. X. Zhong, C. Xiang, M. Yu, Z. Qi, H. Guan, A virtualization based monitoring system for mini-intrusive live forensics. Int. J. Parallel Prog. **43**(3), 455–471 (2015)
17. S. Ahmad, S. Ahmad, B. Li, Honeypot, network forensic technique and framework, in *The 2014 International Conference on Computer Science and Network Security*, (Xi'an, China, 2014), 12–13 April 2014, pp. 407–411
18. D. Srinivasan, X. Jiang, Time-traveling forensic analysis of vm-based high-interaction honeypots, in *Security and Privacy in Communication Networks* (Springer, 2012), pp. 209–226
19. F. Raynal, Y. Berthier, P. Biondi, D. Kaminsky, Honeypot forensics, in *Proceedings from the Fifth Annual IEEE SMC Information Assurance Workshop* (IEEE, 2004), pp. 22–29
20. A. Yasinsac, Y. Manzano, Honeytraps, a network forensic tool, in *Sixth Multi-Conference on Systemics, Cybernetics and Informatics* (2002)
21. B. Cusack, Botnet forensic investigation techniques and cost evaluation, in *Proceedings of the Conference on Digital Forensics, Security and Law* (2014), pp. 171–190
22. S. Krasser, G. Conti, J. Grizzard, J. Gribschaw, H. Owen, Real-time and forensic network data analysis using animated and coordinated visualization, in *Proceedings from the Sixth Annual IEEE SMC Information Assurance Workshop, 2005 IAW'05* (IEEE, 2005), pp. 42–49
23. V.-H. Pham, M. Dacier, Honeypot trace forensics: the observation viewpoint matters. Future Gener. Comput. Syst. **27**(5), 539–546 (2011)
24. A. Pawar, K. Siddhabhati, S. Bhise, S. Tamhane, Web application honeypot. Int. J. **2**(3) (2014)
25. L. Spitzner, The value of honeypots, part one: definitions and values of honeypots (2001)

Chapter 6
Honeypot Framework, Limitation and Counter-Measure

Abstract This chapter describes the conceptual framework of the honeypot. A detailed diagram reveals the workflow of a honeypot. Honeypot does have limitation and that is its common features which can be easily speculated by the attackers. The technique used will be discussed in detail. The distinct difference between the honeypot and a real system is if it gives away its identity. In order for the Research community to stay ahead of this cat and mouse game, we here in the chapter review many honeypot detecting methods proposed by recent researchers so as to improve some of the features. /the online available honeypot detection tool is referenced in this chapter as well.

6.1 Conceptual Framework

Honeypot has been used by many security researchers as a primary tool to learn and understand the behavior of malware and cyber-attack. It provides rich and relevance information of its attacker or malware.

The honeypot framework can be conceptualized into three parts, they are input, functionality and output.

In the input part, cyber-attack is anticipated. The attack can come in a form of network traffic packet such as distributed denial of service [1]. The input can also be in other form such as malware executable file, spam mail, code injected into other popular multimedia or document format [2], driveby download [3], advanced persistent attack [4] or break in by insider [5].

The input of the attack can be classified into two categories, namely planned and unplanned. The planned attacks such as DDoS, APT and insider attack have a specified target and the unplanned attack such as spam, attack caused by driveby download, code injection are basically caused by ignorance user.

Variety of frameworks have been proposed by researcher to monitor and study or to absorb the attacks so as to divert from their primary objective. This will be explained in the functionality part. The frameworks include roaming high-interactive honeypot to prevent DDoS attack [6], service-emulation low interactive honeypot for absorb

© The Author(s) 2018

C. K. NG et al., *Honeypot Frameworks and their Applications: A New Framework*, SpringerBriefs on Cyber Security Systems and Networks, https://doi.org/10.1007/978-981-10-7739-5_6

attack on server [7], virtual high interactive honeypot for client side attack, honeynet to study the behavior of worm. Figure 6.1 is a collection of concepts available in the honeypot framework. Difference concept is introduced due to its capabilities to tackle a very specify problem. How do we know that there is an attack take place in the honeypot? Most honeypot framework consist of an IDS, honeypot/s and a server. The IDS is used to notify its creator of an intrusion. It can be used as a traffic controller to redirect malicious traffic to the honeypot, depend on where it is setup. The honeypots, as mentioned earlier, is not a dummy computer that sit there to be compromised. In fact, honeypot is a computer with the most state-of-the-art monitoring, tracking and communication mechanism. The mechanisms of honeypot consist of tracking, alerting, monitoring and communication function (Fig. 6.2).

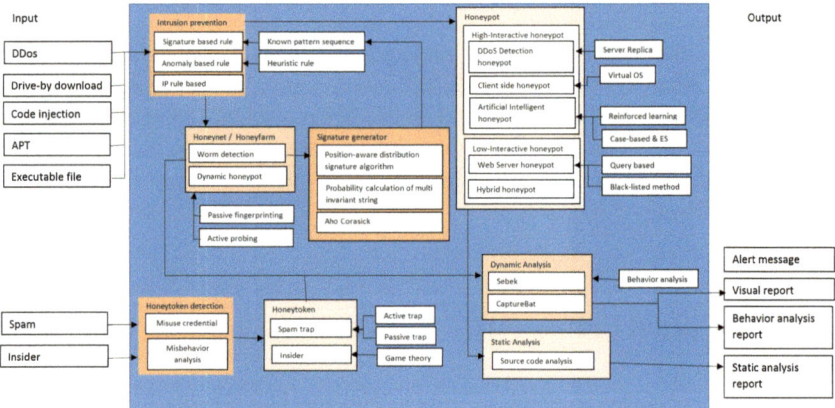

Fig. 6.1 Conceptual honeypot framework

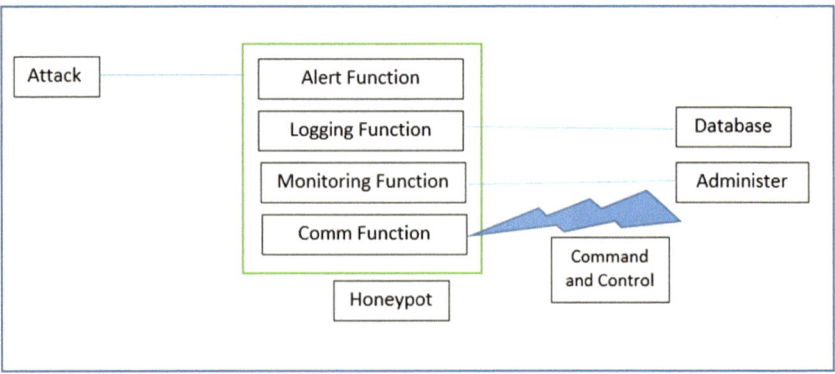

Fig. 6.2 The framework of honeypot

1. Tracking function records every movement executed by its intruder. The tracking mechanism such as Sebek and CaptureBat is hooked to the kernel to prevent discovery.
2. Honeypot can have alerting function if IDS is implemented in it.
3. Monitoring function allows administrator to monitor/ view the activity in the honeypot
4. Communication function provides a pathway for the system administrator to execute command and control the honeypot. Most high interactive honeypots are connected to two networks. One is visible network which is connected to the router and another is the stealth network which is established for communication between the creator and the honeypot. This network is also used to transfer data to the server.

Honeypot can have other exclusive function depend on its purpose. Example, AI honeypot have the ability to interact with its human attacker via using machine learning technique.

Honeypot allows the intruder to mingle around, this enables researcher to learn about the intruder. This includes surfing habit, objective, technical skill and knowledge of its target.

Honeypot provides data to the people such as the system administrator or security researcher about the break-in. At the point of break-in, the honeypot will notify its creator by means of email or even text message. Any activities associated with the intrusion is logged/ recorded and send to a server. Output of the honeypot can also be presented visually so as to allow its creator to have a first-hand account of the on-going event in the honeypot [1].

Server is responsible for storing the data collected by the honeypot. Most servers possess the ability to analyze the static and dynamic data so as to transform them into useful information.

6.2 Common Features of Honeypot

Some of the common features of honeypot have been noted over the year. The knowledge of detecting honeypot has grown among the blackhat community; more sophisticated technique has been introduced and used to determine honeypot.

It is important to know the level of knowledge the public at large possessed about the honeypot and the amount of information about honeypot available online to the public. Acknowledging and appreciating the amount of effort contribute in this area by the blackhat community is importance so as to encourage much deeper thinking to improve the technology and methodology use in the honeypot.

In one of the hacker web blogs, the blogger has listed out a number of basic features of honeypot without using special tool. Below listed some features:

1. The age-old adage, "if it is too good to be true, it probably is", applies as well to hacking. Those sites that seem extraordinarily easy to hack are likely traps.

2. Look for unusual services and ports open. Most internet-facing systems are stripped of any unnecessary services. If it has lot of unusual services and ports open, these are meant to attract attackers and it may be a honeypot.
3. If it is a default install, it may be a honeypot.
4. If there is little or no activity, it may be a honeypot.
5. If you see directories with names such a "social security numbers" or "credit card numbers", it may be a honeypot.
6. If you see very little software installed, it may be a honeypot.
7. If there is a lot of free space on the hard drive, it may be a honeypot.

These seven points mentioned above are consistent with the features of honeypot suggested by S. Mukkamala in his article "Detection of virtual environments and low interaction honeypot" [8]. Mukkamala compares the feature of the honeypot and the benign production system. He pointed out numerous "features" possess by the honeypot which actual system does not have.

1. Honeypot have not network activities in its network
2. High software overhead.
3. All interaction is logged.
4. Bandwidth restriction to prevent damaging done to other system.

The above points give a simple general overview of what to lookout for in the basic honeypot setup theoretically. In reality, different level of interactive honeypot does vary in features. The virtual honeypot has different feature to physical honeypot and high-interactive honeypot holds a different set of feature than its counterpart. *Thorsten Holz* has reveal method to detect virtual high-interaction honeypot. Honeypot like Sebek can be detected and disable by restoring the service descriptor table in the operating system. Shell script like Kebes is used to avoid the logging mechanism [9].

The distinction between virtual honeypot and the real system allows the advanced attacker to tell if the system he/ she attacks is a virtually created high-interactive honeypot by observe following points: for UML, Identify the main thread and other thread. For VMware, check the hardware list, Mac address of the device and I/O backdoor. Virtual honeypot can be easily detected by observing its long execution and response time.

The detection of low-interactive honeypot requires a different approach to detect the virtual high-interactive honeypot. Mukkamala states the characteristic of virtual environment and the virtue of low-interactive honeypot. Three methods can be used to determine low-interactive honeypot and its environment [8]. They are service exercising, timing analysis of ICMP ECHO requests and TCP/IP finger printing. Low-interactive honeypot emulates a limited set of common service. Service exercising involves executing uncommon operations on the compromised system. In timing analysis of ICMP ECHO requests, the central of the assumption on slower rate of ICMP respond in the emulated environment is significance and this assumption is consistent with argument made by Thorsten Holz on detecting the virtual environment. The attacker is able to conclude the nature of the network environment by observing the timing of the respond. In TCP/IP finger printing, active fingerprinting is used to collect the data for analysis. The attacker can analyze the features extracted and compare with the legitimate system.

Low-interactive honeypot can be detected by examine its response timing after a packet has been sent. *Xinwen Fu* has pointed out the response issue of the low-interactive virtual honeypot. The honeypot like honeyd can easily be detected by examine the link latency within the network emulated [10]. Despite the different term is being used, the author is referring to the response rate. The virtual honeypot is not design to emulate the temporal behavior of nodes at high fidelity and it display a different temporal signature to the real physical system and network.

6.3   Other Honeypot Detection Method

Honeypot can also be detected by knowing its legal obligation. *Cliff and Ping Wang* outlined a number of detection methods which are implemented by the bot master via taking advantage of the outgoing restriction posed on the honeypot by the law to detect honeypot.

The detection method is being classified into two different types, namely detection through infection and detection through other illicit activities [11]. In the latter method, the bot master will let the compromised machine send an infectious traffic to the sensor which is randomly selected from the bot pool. The attacker will check to ensure that the outgoing traffic is not modified. This poses a tremendous challenge to the administrator as he/she cannot block or modify the outgoing traffic. In the former method, illicit activities such as low rate port scanning, web requests and email spamming can be used to determine whether the targeted system is a honeypot. The author has also provide detailed information about and how a peer-to-peer botnet based on "buddy list" can be the preferred botnet communication topology used by the bot master.

The bot master in peer-to-peer botnet using multiple bot controllers as sensor to accept the response from the newly created bot in order to ensure that it is not a honeypot before accepting it into the botnet [12]. Bot master will hard code bot controllers' domain name rather than their IP address in all bots and try to keep their bot controller mobile by using dynamic DNS. This is to provide flexibility once it is detected that the bot controller is a honeypot. The bot master can create another bot controller on a new compromised machine and update the DDNS entry to point to the new bot controller.

The features discussed can be used as a scale to determine whether the compromised system is a honeypot. Researcher such as Asama Hayatle proposes the use of Dempster-Shafer theory to detect honeypot [13]. The theory gives a high accurate rating in detection of honeypot. The theory uses the numerous evident to calculate the probabilities of honeypot. Each evident is weighted and BBA (basic belief assignment) value for each evident is determined. The end-result is then compared with the normal system threshold and the honeypot threshold. Other features such as software diversity and activities can be used to determine whether the compromised system is a honeypot.

Honeypot can also be discovered using honeypot detection application which is available freely online to the public at large. Majority of such applications are created by the freelance hacker.

Bill McCarty evaluated the effectiveness of various version of one of the free tool, honeypot hunter. Honeypot hunter was released right after honeypot was introduced by Lance Spaitzer. Over the years, other tools such as THC-Amap, Hping and Nessus gradually become available for the black hat community to be used to detect honeypot [14, 15]. Most of the tools exploit the limitation of the low-interactive honeypot such as low-interactive honeypot's inability to complete the proper handshake process.

The methods used to detect honeypot do come with setback which attacker needs to consider. For timing analysis of ICMP ECHO requests, the long respond time can be contribute by the distance between the victim and the attacker and not always necessary a sign for virtual honeypot. The longer distance implies that more hop the packet has to pass through before reaching the victim machine [8]. Attacker has overcame such problem by extracting the timing of respond of the other host within the same network. Passive network sniffing can be used to resolve the above mentioned problem. This also resolves the problem posed by the latter solution which is how the attacker can be confident that timing extracted from another host within the network is from the legitimate system. The information collected by the sniffing provides the respond timing for each host.

Researcher such as *Xinwen Fu* solves this honeypot link latency issue or timing issue by re-adjust the link latency of the virtual honeypot in the order of one microsecond rather than one millisecond [10]. However, this create another issue. The consistent RTT of the link latency may rise the eyebrow of the attacker to suspect that the victim is a virtually created honeypot. Ping based approach, TCP based approach and UDP based approach are added into the honeypot to create a more realistic RTT to deceive the advanced attacker.

The proposal of a honeypot system almost always comprises of some form of intrusion detection system or firewall to differential and redirect the malicious traffic from the legitimate packet. In order to discover the honeypot without being discovered by or redirected to the honeypot, it is necessary for the attacker to slip through the eye of the security system.

Intrusion Detection System can be evaded by changing the traffic. This involves using a different protocol such as UDP instead of TCP or http instead of ICMP. Additional method such as session splicing can be used to break an attack up into small packet, transmits, reassembles and compromises the host machine. Other methods such as extra data, obfuscating addresses and data by using encryption can also be implement to evade the IDS. It is still possible to detect such attack by setting the anomaly IDS to be more sensitive, however this poses another problem. Some of the legitimate traffic may be classified as malicious threat which result in high false negative. It is advisable to implement more than one anomaly IDS at a single point of check to improve the false negative rate.

To bypass the check from firewall, tool such as 007 Shell can be used to creates a covert tunnel and makes the attack look innocuous to the firewall. This method makes detection virtually impossible for the system administrator. Researcher has formulated tools such as spyhunter which can be used to detect 007 shell.

Hacker and researcher have proposed sever methods and tool to detect honeypot for a different purpose. The aim of hacker is to prevent being discovered so that he/she can further inflict the system to gain advantage. The researcher on the other hand is to discover the drawback of the honeypot so that improvement can be done to make honeypot look like the legitimate host. Method and tools to detect honeypot are continuously evolved and improved by the blackhat community; and the researchers specialized in this field will continuously proposed new idea and concept to improve the concealment and the features of the honeypot to make it difficult to distinguish between the legitimate and fake host. It is for sure that this "discovered and improve" game will continue infinitely.

References

1. S.S. Sadamate, V. Nandedkar, in *Advance Honeypot Mechanism-the Hybrid Solution for Enhancing Computer System Security with DoS*, vol. 4 (2015)
2. M. Akiyama, Y. Kawakoya, T. Hariu, Scalable and performance-efficient client honeypot on high interaction system, in *2012 IEEE/IPSJ 12th International Symposium on Applications and the Internet (SAINT)* (IEEE, 2012), pp. 40–50
3. T.-M. Koo, H.-C. Chang, Y.-T. Hsu, and H.-Y. Lin, Malicious website detection based on honeypot systems, in *2nd International Conference on Advances in Computer Science and Engineering (CSE 2013)* (Atlantis Press, 2013)
4. P. Chen, L. Desmet, C. Huygens, A study on advanced persistent threats, in *Communications and Multimedia Security* (Springer, 2014), pp. 63–72
5. B.M. Bowen, M.B. Salem, A.D. Keromytis, S.J. Stolfo, Monitoring technologies for mitigating insider threats, in *Insider Threats in Cyber Security* (Springer, 2010), pp. 197–217
6. S. M. Khattab, C. Sangpachatanaruk, D. Mossé, R. Melhem, T. Znati, Roaming honeypots for mitigating service-level denial-of-service attacks, in *24th International Conference on Distributed Computing Systems, 2004. Proceedings* (IEEE, 2004), pp. 328–337
7. J.P. John, F. Yu, Y. Xie, A. Krishnamurthy, M. Abadi, Heat-seeking honeypots: design and experience, in *Proceedings of the 20th International Conference on World Wide Web* (ACM, 2011), pp. 207–216
8. S. Mukkamala, K. Yendrapalli, R. Basnet, M. Shankarapani, A. Sung, Detection of virtual environments and low interaction honeypots, in *Information Assurance and Security Workshop, 2007. IAW'07. IEEE SMC* (IEEE, 2007), pp. 92–98
9. T. Holz, F. Raynal, Detecting honeypots and other suspicious environments, in *Proceedings from the Sixth Annual IEEE SMC on Information Assurance Workshop, 2005. IAW'05* (IEEE, 2005), pp. 29–36
10. X. Fu, W. Yu, D. Cheng, X. Tan, K. Streff, S. Graham, On recognizing virtual honeypots and countermeasures, in 2nd *IEEE International Symposium on Dependable, Autonomic and Secure Computing* (IEEE, 2006), pp. 211–218
11. C.C. Zou, R. Cunningham, Honeypot-aware advanced botnet construction and maintenance, in *International Conference on Dependable Systems and Networks, 2006. DSN 2006* (IEEE, 2006), pp. 199–208
12. P. Wang, L. Wu, R. Cunningham, C.C. Zou, Honeypot detection in advanced botnet attacks. Int. J. Inf. Comput. Secur. **4**(1), 30–51 (2010)

13. O. Hayatle, A. Youssef, H. Otrok, Dempster-shafer evidence combining for (anti)-honeypot technologies. Inf. Secur. J. Glob. Perspect. **21**(6), 306–316 (2012)
14. K. Graves, in *CEH Certified Ethical Hacker Study Guide* (Wiley, 2010)
15. M. Gregg, in *Certified Ethical Hacker (CEH) Cert Guide* (Pearson IT Certification, 2013)

Chapter 7
Ramsonware and Honeypot

Abstract Ramsonware is a malware of its kind and is growing rapidly. There is an urgency to take this matter seriously. Honeypot which very few focus on is an excellence tool to gather information of ransomware. To deal with ransomware, it is important to understand the payment system, bitcoin. The proposal of using honeytoken to track the location of the receiver of bitcoin can help find the root of the cause.

7.1 Ransomware

Ransomware is a form of malware that encrypt important files or disable the certain function of the computer [1]. It also provides service for decrypting files or unlock terminal with the exchange of online currency such as bitcoin or moneypak.

Ramsonware has become more and more popular among the blackhat community over the past decade with the entry of the cryptocurrency in 2009 such as Bitcoin. Most ransomware variances such as cryptolocker, cryptowall and torrentlocker demand bitcoin from their victim as a form of payment. The cryptocurrency allows ramsonware user to stay anonymous.

The tasks of ransomware can be basically classified into three different phases. They are targeted phase, encrypted phase and messaging phase. All cryptolocker ransomwares follow this principle.

Ransomware can be spread using the following tactic reveal below [2]:

1. Embedding code in popular file extension such as word document, video and audio file
2. Old fashion driveby-download method from compromised website
3. Downloader and Botnets
4. Spam email
5. Social engineering and Self-propagation
6. Malvertisement

© The Author(s) 2018   75
C. K. NG et al., *Honeypot Frameworks and their Applications:*
A New Framework, SpringerBriefs on Cyber Security Systems and Networks,
https://doi.org/10.1007/978-981-10-7739-5_7

Ransomware has moved from a random attack to a more target specific attack. It can be seen in the recent outbreak in the hospitals in United States and Canada by samsam ransomware [3]. Unlike most ransowmares, Samsam does not spread through any of the method mentioned above, instead, it uses APT technique to attack the targeted system [4]. The attacker monitor, collect and analysis the vulnerability of server before attacking them.

The innovation of the ransomware has included the self-propagation technique used in zcrypt and sophisticated multiple layer encryption used in the anti-detection technique by Cerber ransomware. The new version of the same ransomware class will supersede the older version by having a better encryption technique, and some may include more sophisticated anti-detection and anti-debugging technique to replace the old technique.

7.2   Ransomware Honeypot

Article on anti-ramsonware concept and cryptocurrency have made their way into some of the major conferences. Researchers have proposed several method to detect ransomware and very few studies ransomware in honeypot perspective.

Krzysztof Cabaj used honeypot and a fake dns server to analyse cryptowall ransomware so as to understand its work pattern and communication channel [5]. *Chris Moore* proposed the use of honeypot technique to detect ransomware. The detection is determined using hierarchy of responses [6]. *Amin Kharraz* used honeypot to house the redirected traffic to prevent the malicious traffic to cause damage to host outside the network [6]. References [5–7] use single honeypot to collect data or detect ransomware.

The proposed framework uses multiple honeypot to detect ransomware. The primary honeypot consists of a single active client-based honeypot. This honeypot with custom static and dynamic analysis engine can be implemented to capture and study the activity of ransomware attack [8]. The system logs all actions performed by the ramsonware and also the information of the key used to encrypt the document [9].

The secondary honeypots consists of multiple passive server-based honeypots with the aid of remunx framework [10] are included into the proposed architecture. The addition of such honeypot helps to capture the propagation of ransomware if any. The server-based honeypot able to capture target-oriented ransomware like Samsam class type. Fake DNS server will be setup to expand all the hardcoded server locations where the CNC or the encryption key is stored [5].

Most ransomwares use automated malware script which they will communicate with the server to get information or send information. Most of the ransomwares require their victim to use Tor browser to communicate with the server [11], this is to keep secret of their location or the server location. Honeypots are capable to capture useful information such as the unique pattern of how each ransomware work. The proposed framework can actively engage ransamware simple or malicious web site to collect information or be used as a passive honeypot waiting for ransomware to attack. This framework helps to expose the limitation and other capabilities of the ransomware. The data collected can be use to generate a new signature for the string-based intrusion detection system or create a pattern data for anomaly-based intrusion detection system.

7.3 Bitcoin Honeytoken

In order to locate the origination of the ransomware, it is necessary to setup a payment trap called the bitcoin honeypot. The setup of the cryptocurrency honeypot does face challenges and questions while considering various honeypot concept for the task. They are:

1. Cryptocurrency does not have the destination IP information. The currency is transferred based on the currency address which is a public key in a P2P network.
2. What is the level of probability of the success rate to locate the receiver?
3. What is the probability rate that the track down IP address belong to the receiver and not the peer?

The framework proposed by [12] for his insider detection honeypot should be considered. The bitcoin honeypot will be an hybrid honeypot which acts as a server that constantly listening for the returning signal sent by the honeytoken. It is also client honeypot that sends "payment" to be traced. The honeytoken is embedded into the bitcoin transaction so that once the receiver receive the payment, it will trigger the honeytoken to send signal back to the sensor. The signal received is an ICMP packet which will reveal the source IP address.

References

1. A. Gazet, Comparative analysis of various ransomware virii. J. Comput. Virol. **6**(1), 77–90 (2010)
2. M.H.U. Salvi and M.R.V. Kerkar, Ransomware: A cyber extortion (2016)
3. D.F. Sittig, H. Singh, A socio-technical approach to preventing, mitigating, and recovering from ransomware attacks. Appl. Clin. Inf. **7**(2), 624 (2016)
4. S.S. Response, Samsam may signal a new trend of targeted ransomware (2016), https://www.symantec.com/connect/blogs/samsam-may-signal-new-trend-targeted-ransomware
5. K. Cabaj, P. Gawkowski, K. Grochowski, D. Osojca, Network activity analysis of cryptowall ransomware. Przeglad Elektrotechniczny **91**(11), 201–204 (2015)
6. C. Moore, Detecting ransomware with honeypot techniques, in *Cybersecurity and Cyberforensics Conference (CCC), 2016* (IEEE, 2016), pp. 77–81
7. A. Kharraz, W. Robertson, D. Balzarotti, L. Bilge, E. Kirda, Cutting the gordian knot: a look under the hood of ransomware attacks, in *International Conference on Detection of Intrusions and Malware, and Vulnerability Assessment* (Springer, 2015), pp. 3–24
8. T.-M. Koo, H.-C. Chang, Y.-T. Hsu, H.-Y. Lin, Malicious website detection based on honeypot systems, in *2nd International Conference on Advances in Computer Science and Engineering (CSE 2013)* (Atlantis Press, 2013)

9. C. Seifert, R. Steenson, I. Welch, P. Komisarczuk, B. Endicott-Popovsky, Capture-a behavioral analysis tool for applications and documents. Digit. Investig. **4**(Supplement), 23–30 (2007)
10. L. Pearce, Malware analysis in a nutshell. Technical Report (Los Alamos National Laboratory (LANL), 2016)
11. D. McCoy, K. Bauer, D. Grunwald, T. Kohno, D. Sicker, Shining light in dark places: understanding the tor network, in *International Symposium on Privacy Enhancing Technologies Symposium* (Springer, 2008), pp. 63–76
12. B.M. Bowen, M.B. Salem, A.D. Keromytis, S.J. Stolfo, Monitoring technologies for mitigating insider threats, in *Insider Threats in Cyber Security* (Springer, 2010), pp. 197–217

Chapter 8
Conclusions and Future Work

Abstract The last chapter is comprised of 2 sections. First section gives a brief history of honeypot and re-visits the framework and concept of honeypot. The second section emphasizes on the beneficiary of honeypot in the perspective of detecting ransomware and its future benefits.

8.1 Conclusion

This book discusses the concepts of the honeypot. It also describes and compares different honeypot frameworks for general and specialized attacks.

8.1.1 Summary

The timeline in Fig. 8.1 reveals that the recent development of the honeypot has moved in the following direction technologically:

1. Dynamic Honeypot
2. Artificial Intelligent Honeypot

Despite the growth of popularity in the highly sophisticated honeypot, concept like shadow honeypot which is introduced in the early 2000 has diminished in its light of focus. This is due to the combined use of honeypot and anomaly IDS has become a norm in almost all architecture proposed by researcher. But it is worthwhile to include and discuss its development to show appreciation for its contribution in the area of network security.

The threats that have been discussed in this survey are summarized below:

1. Worm
2. Phishing and spam email
3. APT
4. DDoS
5. Insider Threat

© The Author(s) 2018
C. K. NG et al., *Honeypot Frameworks and their Applications:
A New Framework*, SpringerBriefs on Cyber Security Systems and Networks,
https://doi.org/10.1007/978-981-10-7739-5_8

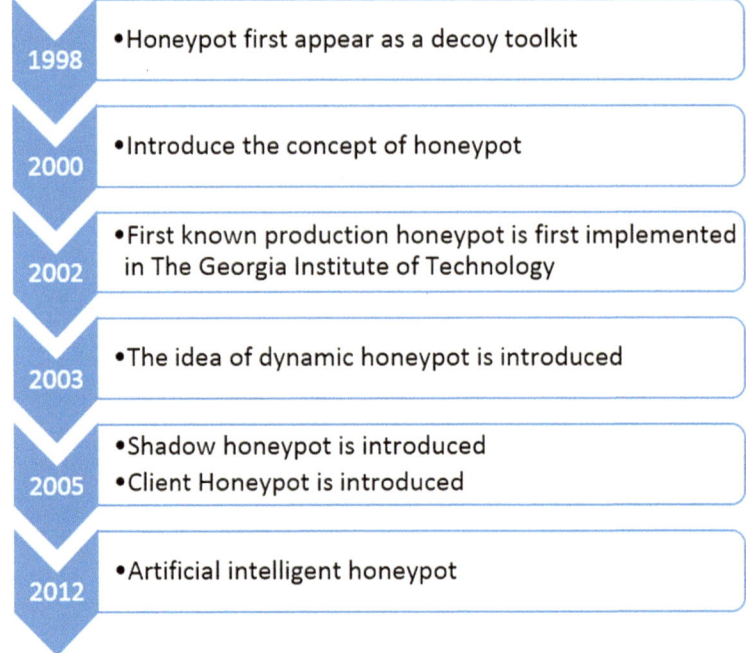

Fig. 8.1 Timeline for the development of Honeypot

The purpose of honeypot determines its architecture design and framework. An example of a worm detection honeypot, requires two honeynets to be placed side by side for the worm to roam between and compromise them in order to get a full set of its variants. The bot detection honeypot may only require one or two honeypots at most to capture and identify the botnet. Honeypots are a good tool for research. It provides a sandbox-like concept to contain the attack. Honeypots are also a useful tool (not security tool) for security purpose. It is integrated as part of the security system.

One shortcoming of honeypots is for example the limited features in low-interactive honeypot and the response timing for virtual honeypot, have been covered extensively and researcher has proposed solution to overcome or enhance the problems.

Despite the benefit of honeypots described above, they do face other challenges which have not been clearly addressed. One challenge is the limitation in the proposed honeypot architecture for detection of APT. The proposed architecture fails to cover all possible attack which can be part of APT attack. One example is that, the architectural design needs to consider external attacks such as those initiated by spam mail (outsider attack).

Honeypots are a excellence supporting tools for security, though they should not be used to replace any security tools of any organisation.

8.1.2 Future Research Work

The rise in popularity of ramsonware has sounded the alarm and immediate attention is necessary to look into this issue. As mentioned, ramsonware detection based on honeypot concept is not receive much attention and it is perceived to have great potential for honeypot to be use in such direction. Honeypot can be used to collect useful information about ransomware especially the crypto algorithm used to encrypt the file. The concept of honeytoken can be used to trace Bitcoin transaction destination by emitting signal embedded in the transaction to send back to the sender.